Robert M. Koerner and Joseph P. Welsh
CONSTRUCTION AND GEOTECHNICAL ENGINEERING
USING SYNTHETIC FABRICS

J. Patrick Powers
CONSTRUCTION DEWATERING: A GUIDE TO THEORY AND
PRACTICE

Harold J. Rosen
CONSTRUCTION SPECIFICATIONS WRITING:
PRINCIPLES AND PROCEDURES, Second Edition

Walter Podolny, Jr. and Jean M. Müller
CONSTRUCTION AND DESIGN OF PRESTRESSED
CONCRETE SEGMENTAL BRIDGES

Ben C. Gerwick, Jr. and John C. Woolery
CONSTRUCTION AND ENGINEERING MARKETING
FOR MAJOR PROJECT SERVICES

James E. Clyde
CONSTRUCTION INSPECTION: A FIELD GUIDE TO PRACTICE,
Second Edition

Julian R. Panek and John Philip Cook
CONSTRUCTION SEALANTS AND ADHESIVES, Second Edition

Courtland A. Collier and Don A. Halperin
CONSTRUCTION FUNDING: WHERE THE MONEY COMES
FROM, Second Edition

James B. Fullman
CONSTRUCTION SAFETY, SECURITY, AND LOSS PREVENTION

Harold J. Rosen
CONSTRUCTION MATERIALS FOR ARCHITECTURE

CONSTRUCTION
MATERIALS
FOR
ARCHITECTURE

Construction Materials For

ARCHITECTURE

HAROLD J. ROSEN, PE, FCSI

Illustrations Drawn by
PETER M. ROSEN, RA

A WILEY-INTERSCIENCE PUBLICATION

JOHN WILEY & SONS

New York Chichester Brisbane Toronto Singapore

The detailed information, drawings, tables, and other
data in this book have been accumulated over the years
by the author from governmental sources, trade
organizations, building materials manufacturers, and other
professional specification writers. The author has made a
reasonable attempt to ascertain the validity of the data
presented herein, but does not warrant, and assumes no
responsibility for the accuracy or the completeness of the
text material, drawings, and other data. The user as in
any other investigation must consult all sources of
information and make a professional judgment based on
all the relevant facts.

Library of Congress Cataloging in Publication Data:

Rosen, Harold J.
   Construction materials for architecture.

   (Wiley series of practical construction guides)
   "A Wiley-Interscience publication."
   Bibliography: p.
   includes index.
   1. Building materials.   2. Architecture.   I. Title.
II. Series.

TA403.6.R67   1985   691   84-19700
ISBN 0-471-86421-8

Printed in the United States of America

10 9 8 7 6 5 4 3 2 1

SERIES PREFACE

The Wiley Series of Practical Construction Guides provides the working constructor with up-to-date information that can help to increase the job profit margin. These guidebooks, which are scaled mainly for practice, but include the necessary theory and design, should aid a construction contractor in approaching work problems with more knowledgeable confidence. The guides should be useful also to engineers, architects, planners, specification writers, project managers, superintendents, materials and equipment manufacturers and, the source of all these callings, instructors and their students.

Construction in the United States alone will reach $250 billion a year in the early 1980s. In all nations, the business of building will continue to grow at a phenomenal rate, because the population proliferation demands new living, working, and recreational facilities. This construction will have to be more substantial, thus demanding a more professional performance from the contractor. Before science and technology had seriously affected the ideas, job plans, financing, and erection of structures, most contractors developed their know-how by field trial-and-error. Wheels, small and large, were constantly being reinvented in all sectors, because there was no interchange of knowledge. The current complexity of construction, even in more rural areas, has revealed a clear need for more proficient, professional methods and tools in both practice and learning.

Because construction is highly competitive, some practical technology is necessarily proprietary. But most practical day-to-day problems are common to the whole construction industry. These are the subjects for the Wiley Practical Construction Guides.

M. D. MORRIS, P.E.

PREFACE

As the author of several books on specification writing, I have constantly reiterated the admonition that a knowledge of specification writing is not in and of itself sufficient for one to be a good specifier. The truly competent specifier requires a broad knowledge of construction materials. However, for the most part this knowledge has not yet been acquired in our colleges and universities.

In addition, the specifier, architect, and selector of construction materials must be able to assess, evaluate, and utilize these materials in a design so that the product performs as intended with the least amount of deterioration, degradation, or failure.

Materials are manufactured that have a wide range of characteristics, and one must select a product having those criteria that will perform, given the economics and the longevity that are essential for the specific project. For products that are manufactured on the job site (i.e. concrete, masonry, roofing, waterproofing, etc.), the basic ingredients comprising the finished product and quality of work involved in the on-site operations must be mastered so that the finished product achieves a high degree of durability.

Paralleling the need for a comprehensive understanding of materials is the requirement for a rational method of evaluating and selecting products for a specific use in a project. The absence of such an investigative tool invites possible construction failure, since some pertinent performance characteristic may have been overlooked during the course of the usual review, unrelated to a programmed evaluation.

Chapter 1, Performance Considerations, sets forth such a rational method of evaluation and selection. This approach had its genesis in a course entitled, "A Systematic Approach to Building Material Evaluation and Selection," which I began directing at the University of Wisconsin—Extension in 1975, under the supervision of Philip M. Bennett, Program Director. In addition to the systematic approach outlined in Chapter 1, several materials are subjected to the same analysis throughout this book so that the reader may understand the process more productively and will be encouraged to utilize the approach in the quest for better solutions.

The materials dealt with in this book, while encompassing the basic materials generally encountered in a project, are primarily those that are produced or assembled on the site, such as architectural concrete, masonry, ornamental metal, architectural woodwork, roofing, waterproofing, curtain walls, and so forth. They all require a knowledge of on-site manufacture to produce the desired end results. In addition, the book presents in detail the utilization of older materials for new, innovative uses, such as concrete for architectural purposes and building stone for thin exterior panels. Newer products and techniques such as single ply roofing, inverted roofing, sealants, curtain walls, modern day paints, and seamless flooring are likewise discussed along with the means to evaluate and assess their specific characteristics.

Having spent some forty years as a specifier, commencing during World War II, and as the author of *Specifications Clinic for Progressive Architecture* for some twenty years, I have accumulated a wealth of information that should be transmitted to those involved in the business of material selection, evaluation, and specifying.

This undertaking has been lightened by the generous cooperation of a number of individuals. My wife, Rose, who typed the manuscript; my son Peter, an architect with the firm of Davis & Carter, McLean, VA, who prepared the illustrations; Gordon Wildermuth, a partner in the firm of Skidmore Owings & Merrill, who permitted me to select photographs from work in which I was involved; Werner Wandelmaier, a partner in the firm of I.M. Pei & Partners, who permitted me to utilize photographs of some of that firm's work; and Maurice Lehv, who photographed a number of pictures from various publications, which appear throughout this book.

HAROLD J. ROSEN

Coconut Creek, Florida
January 1985

CONTENTS

PERFORMANCE CONSIDERATIONS

The performance of building materials, products, components, and assemblies can be comprehended more fully if certain basic information regarding these characteristics is understood.

Building materials and their composites must serve an intended function over a certain life span of the structure. Some materials, such as concrete, masonry, stone, metals, and plaster, will usually last the lifetime of the building. Other materials, such as roofing (built-up bituminous and elastomeric sheets), sealants, and paints, have more limited life expectancies and will require replacement after some period of time. The useful life expectancy of materials is related to the environmental conditions for which they were first selected; their durability or service life is a function of that environment.

Predictions as to the behavior and performance of traditional building materials under given geographical and environmental conditions can be made quite accurately. However, since there has been less use experience with more recently developed building materials, primarily those which are the products of modern day chemistry, metallurgy and technology, the same long-term performance behavior lessons cannot be applied to them.

Indeed, even traditional materials fare differently when used and exposed in new climes such as the harsh environs of the Middle East. Witness also the unusual behavior of stone and copper, which survived centuries of use in Venice and Athens but are now succumbing to the effects of environmental industrial fumes and acid rain.

The performance, or service life, or durability of a material is a function not only of environment (weather, temperature, rain, humidity, ozone, bacterial attack, solar radiation, etc.) but also of physical interaction, both natural (wind loads, seismic forces) and human-made (pollution, physical abuse). The influence of environmental and physical interaction results in failure as a result of chemical degradation, excessive expansion and contraction, unusual wear, and a host of other types of failure.

PERFORMANCE CHARACTERISTICS

The performance level of both traditional and new materials may be assessed more accurately by evaluating certain characteristics when selecting products that must serve their intended function over a period of time, taking into consideration the natural environment and the human-made elements that must be endured.

The performance characteristics to be investigated on a rational basis include the following:

Structural serviceability
Fire safety
Habitability
Durability
Compatibility

Some test procedures may be empirical and others may subject the material to environmental conditions approximating the situation to which the material will be exposed, but testing alone will not necessarily predict the final outcome. Knowledge and analytical evaluation are additional ingredients to be used along with testing in predicting performance.

The evaluation process is enhanced by creating a comprehensive listing of subcategories of the five major performance categories to reduce the likelihood of overlooking certain important performance characteristics when making an evaluation.

STRUCTURAL SERVICEABILITY

The performance characteristic of structural serviceability includes resistance to natural forces such as wind and earthquake; structural adequacy; and physical properties such as strength—including compression, tension, shear, and behavior against impact and indentation.

A more expanded listing, although not necessarily complete, includes the following:

Natural Forces
Wind
Seismic

Strength
Compression
Tension
Shear
Torsion
Modulus of rupture
Indentation
Hardness

Architects generally rely on structural engineers to solve such problems as building live and dead loads, seismic forces, wind pressures, earth and water pressures, compression, tension, torque, and so forth. There are times when the architect must recognize that, in evaluating an architectural product or a component, appropriate structural performance characteristics must be reviewed to ascertain whether the proposed products and their installation details also will withstand certain of these forces.

Curtain walls are subjected to wind loads and the selection of glass sizes and thicknesses, and metal retaining members must be carefully analyzed to ensure structural integrity. When new glass products are introduced, consideration must be given to double glazing, triple glazing, combinations of clear and/or coated glass, and annealing and tempering processes to ensure that the configuration selected will withstand the wind loads.

Roofing materials are oftentimes subjected to wind uplift. Architects must review roofing systems and the working drawing details to verify the structural serviceability of the installation.

FIRE SAFETY

An investigation of fire safety includes the resistance of building materials against the effects of fire. A more comprehensive list of these performance characteristics would include the following:

Fire resistance
Flame spread
Smoke development
Toxicity

While fire safety should include design of structures to reduce and isolate fires, such as compartmentation, smoke shafts, areas of refuge, and so forth, this volume is concerned with an evaluation process of building products so that the required fire safety is evaluated and products selected to withstand and reduce the fire hazard.

Fire Resistance

Fire resistance is the capacity of a material or an assembly of materials to withstand fire or provide protection from it; this is characterized by the ability to confine fire, to continue to perform a structural function, or both.

While a steel beam or a steel column will not burn, its ability to perform a structural function is limited because the temperature rise in the steel to about 1000 to 1200°F will produce structural failure. By enveloping the steel in concrete, masonry, drywall, or spray-on fireproofing, its fire resistance is enhanced as a composite.

Fire endurance is a measure of the elapsed time during which a material or assembly provides fire resistance. ASTM E119 is a test method for determining fire endurance (elapsed time), which is a measure of fire resistance.

When evaluating materials or assemblies for fire endurance performance characteristics, there are several standard test methods as follows:

ASTM E119   Fire Tests of Building Construction and Materials

ASTM E152   Fire Tests of Door Assemblies
ASTM E163   Fire Tests of Window Assemblies
UL 263   Standard Fire Test

Flame Spread

Once a fire has developed, the rate and extent of flame spread are influenced by a number of factors that may vary from one fire situation to another. However, experience and evidence illustrate that the flame spread can be greatly influenced by the flammability of the surface materials or linings.

Since the rate of flame spread is a measure of how quickly fire will spread and develop, there are cogent reasons to select surface or lining materials that have a low flame spread index, or rating. Flame spread index is a numerical designation or classification applied to a building material or composite which is a comparative measure of its ability to resist the spread of flame over its surface. To evaluate the flame spread index, there are several ASTM test methods available, as follows:

ASTM E84   Surface Burning Characteristics of Building Materials
ASTM E162   Surface Flammability of Materials using a Radiant Heat Energy Source
ASTM E286   Surface Flammability of Building Materials using an 8 Foot Tunnel Furnace

ASTM makes this caveat regarding the test methods:

"This standard should be used to measure and describe the properties of materials, products or systems in response to heat and flame under controlled laboratory conditions and should not be used for the description or appraisal of the fire hazard of materials, products, or systems under actual fire conditions."

However, numerous building codes have adopted these test procedures and will accept the flame spread index or classification of a material based on these test methods. Figure 1-1 illustrates the ASTM E84 Test Tunnel.

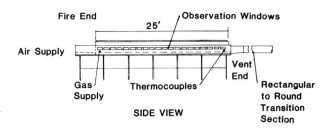

SIDE VIEW

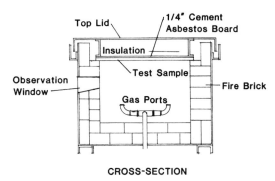

CROSS-SECTION

FIGURE 1-1   ASTM E84 Test Tunnel (Reprinted, with modifications, with permission from ASTM, 1916 Race Street, Phila., PA 19103.)

Smoke Development

The products of combustion include smoke, toxic gases, and vapors along with the flame. Smoke is particulate matter consisting of very fine solid particles and condensed vapor and constitutes most of the visible part of the products of combustion. The main danger from smoke is reduced visibility, which may impede escape from a fire situation and prolong the exposure of occupants to toxic fumes.

The smoke development index or density is a numerical classification based on test method ASTM E84. Some codes may require that surface materials or linings for public exit corridors and stair halls use products with a Class A or low smoke density number.

Toxicity

Gas is one of the products of combustion. Toxic gases such as carbon monoxide, hydrogen cyanide, hydrogen chloride, and nitrogen dioxide are lethal. The Uniform Building Code and the BOCA Basic Building

Code have toxicity requirements. These codes require that the products of combustion for interior finishes on walls and ceilings be no more toxic than those of untreated wood under similar conditions. However, a standard test method that can verify these requirements is lacking.

A test method in use in some areas to determine toxicity includes the heating of the test specimen in a furnace with the decomposition products wafted over to a chamber containing mice. The mice are then examined for lesions, weight loss, and other toxic reactions. The combustion products generated by a specific material under observation must be less toxic than those generated by wood, which is the referenced standard. A recent review entitled *A Critical Review of the State-of-the-Art of Combustion Toxicology*, dated June 1982 by the Southwest Research Institute, examines the various test methods and other studies concerning combustion produced toxic gases.

HABITABILITY

The performance characteristic of habitability include livability as characterized by thermal efficiency, acoustic properties, water permeability, hygiene, comfort, and safety. An expanded version would include the following:

Thermal Properties
Thermal expansion
Thermal transmittance and resistance
Thermal shock

Acoustic Properties
Sound transmission
Sound absorption
Noise reduction coefficient

Water Permeability
Water absorption
Permeability—water vapor transmission
Moisture expansion and drying shrinkage

Hygiene, Comfort, Safety
Toxicity
Vermin infestation
Slip resistance
Mildew resistance
Air infiltration

Thermal Properties

Thermal Expansion

Some buildings exhibit failures relating to thermal expansion and contraction soon after completion or after a significant time frame. These failures include cracking of glass and masonry, failure of sealant joints, heaving of pavement slabs, and so forth. In addition to being unsightly, cracks in exterior surfaces may permit water infiltration, which may result in more serious structural impairment as deterioration proceeds.

Most materials expand with rising temperatures and contract upon cooling. For inorganic solid materials such as masonry, metals, and concrete, the increase in length per unit length for one degree rise in temperature is known as the linear coefficient of thermal expansion. Values for these coefficients of expansion are known and are available in handbooks for design purposes. Organic materials such as sealants and built-up roofing are elastic and move with temperature changes but do not have a memory and do not necessarily revert to their original position or configuration after temperature cycling.

Thermal Transmittance and Resistance

Heat transfer through a material or an assembly of various materials occurs as a result of heat flow from the warmer side to the cooler side. Such heat transfer is proportional to the temperature difference between the two surfaces and further depends upon the specific material or materials and their thickness.

Heat resistance is a measure of the impedance to the flow of heat that a material or an assembly of materials offers.

The American Society of Heating, Refrigerating and Air Conditioning Engineers has defined a series of various coefficients of heat transfer, and ASTM likewise defines under its various standards certain heat transfer terms that are pertinent to their standards. Some of the terms that are frequently associated with building materials are rephrased here. To that extent they may not be as precise, but perhaps they are less obscure.

Thermal Conductance—C or C-value. The time rate at which heat flows through 1 square foot of a material of known thickness in 1 hour when the temperature difference between the surfaces is 1°F. Thermal conductance is expressed as Btu per (hour) (square foot) (°F).

Thermal Conductivity—K or K-value. The time rate at which heat flows through a homogeneous material, 1 inch thick, 1 square foot in area, in 1 hour when the temperature difference between the surfaces is 1°F. Thermal conductivity is expressed as Btu per (hour) (square foot) (°F per inch).

Thermal Resistance—R, R-factor, or R-value. The reciprocal of a heat transfer coefficient (R = 1/C or 1/U). Thermal resistance is expressed in Btu per hour per square foot per °F.

Thermal Transmittance—U or U-value. The time rate of heat flowing through 1 square foot of material, in 1 hour, when the temperature difference between the surfaces is 1°F. Sometimes called the "Overall Coefficient of Heat Transfer." Thermal transmittance is expressed as Btu per (hour) (square foot) (°F).

The most useful coefficient or factor in thermal design calculations is the R-factor, because resistances are additive whereas conductances and conductivities are not additive. R-values are particularly useful for estimating the effects of components of various materials of a system because they can be directly added.

The U-value of a component assembly can be determined by ASTM C236, Test for Thermal Conductance and Transmittance of Built-Up Sections by Means of the Guarded Hot Box. However, the rational method of estimating the U-value by calculation may be determined by adding the resistances R and dividing into 1, thus

$$U = \frac{1}{R_1 + R_2 + R_3 + \ldots}$$

Thermal Shock

The phenomenon of thermal shock is exhibited when a sudden stress is produced in a material as a result of a sudden temperature change. As an example, in some climates or geographical areas, materials (especially dark surfaces) may be exposed to high temperatures of the sun. A sudden rain storm may cause a marked decrease in temperature in a short time interval, producing a thermal shock in the material. Colored glass, brick, and ceramic glazes may exhibit thermal shock under these circumstances.

Acoustic Properties

Acoustics is the science of heard sound, including the propagation, transmission, and effects of sound. Sounds are an integral and necessary or unnecessary part of our daily lives. Those that are necessary (speech communication, music, etc.) are desirable and wanted. Those that are unnecessary (traffic, machinery, jet engines) or which audibly interfere with what we are trying to hear are unwanted sounds, which are generally referred to as noise regardless of their character.

Sound Transmission

Sound transmission is the passage of sound from one space to another as between rooms (through walls, floors, ceilings, doors, partitions, etc.) or from an exterior source to an inside space (through walls, windows, roofs, etc.).

To reduce sound transmission one must consider the sound-isolating properties of the intervening construction, whether it is a wall, partition, floor, or ceiling. This sound-isolating property is the ability of the intervening medium to dissipate significant amounts of sound energy. The sound-reducing capability of the intervening construction is determined by its sound transmission loss (TL), which is the reduction in the sound pressure level usually expressed in decibels.

A construction that transmits only a small amount of noise will have a high sound transmission loss. Conversely, a construction that is paper thin will transmit most of the noise generated and will have a negligible sound transmission loss.

A standard test method and laboratory procedure for the measurement of airborne sound transmission losses is ASTM E90. The transmission loss value measured by this test method results in a single number rating or value called the sound transmission class (STC). Table 1-1 lists some common construction STC values for walls and partitions.

Sound Absorption

Sound absorption is a measure of the property of a material or a construction to absorb sound energy. Sound-absorbing materials have the ability to absorb sound to one degree or another and reflect back a percentage of that sound. They serve as a medium to absorb appreciable amounts of sound energy.

Typical sound-absorbing materials are carpets; acoustical tile; and sound-absorbent products fabri-

TABLE 1-1
AIRBORNE STC VALUES FOR CERTAIN WALLS AND PARTITIONS

Partition or Wall Construction	STC Value
3 in. concrete	47
12 in. brick	56
3 in. cinder block with ⅝ in. plaster both sides	45
6 in. hollow concrete masonry unit	43
12 in. hollow concrete masonry unit	48
2 x 4 in. wood studs, 16 in. O.C., ½ in. gypsum board, both faces	32
Stud wall as above with 2 in. insulation	35
2 x 4 in. wood studs, 16 in. O.C. with ⅜ in. gypsum lath and ½ in. plaster, both faces	46
3 ⅝ in. metal studs, 24 in. O.C. with ⅝ in. gypsum board, both faces	41
3 ⅝ in. metal studs, 24 in. O.C. with double layer ⅝ in. gypsum board, both faces	47

cated of porous materials such as fibrous or cellular plastic.

The sound-absorptive quality of a material or a construction (fibrous materials with cloth or perforated facings) is a function of the incident sound to the reflected sound expressed as a ratio varying from 0 (no absorption) to 1.0 (perfect absorption). Test Method ASTM C423, Sound Absorption and Sound Absorption Coefficients, describes a method of testing and reporting the average sound absorption characteristics of a room, an object such as a screen, and the absorption coefficient of a specimen of sound-absorptive material such as acoustical ceiling tile.

Noise Reduction Coefficient

For a sound-absorptive material and similarly for a hard surface (e.g., plaster, concrete block), a single number rating called the noise reduction coefficient (NRC) is reported. The NRC is the average of the sound absorption coefficients of a material tested at 250, 500, 1000, and 2000 Hz, rounded to the nearest multiple of 0.05. Generally, materials with an NRC of 0.50 provide 50% sound absorption and 50% sound reflection. As the percentage increases to 1.0, the material becomes more efficient as a sound absorber. As the NRC approaches 0, the material tested becomes a reflector of sound. Table 1-2 illustrates some typical materials and their NRC values.

TABLE 1-2
NRC COEFFICIENTS FOR CERTAIN BUILDING MATERIALS

Material	NRC
Brick	0.05
Carpet (no underlayment)	0.45
Carpet (40 ounces underlayment)	0.55
Concrete masonry units painted	0.05
Resilient flooring on concrete	0.05
Glass	0.05
Plaster on lath	0.05
Plywood paneling	0.15
Acoustical title	0.55 to 0.85, depending on type and mounting

FIGURE 1-2 Loss of paint adhesion due to moisture vapor buildup within building.

Water Permeability

Some building materials and constructions are affected by water, which exists in three phases: solid—*ice*; liquid—*water*; gaseous—*water vapor*. Many buildings develop material failures as a result of water entering into their systems.

For example, when water enters exterior concrete, masonry, or similar construction and subsequently freezes to ice, cracking may occur because of the volume increase of water changing from the liquid to the solid state. When gaseous water vapor is trapped below built-up bituminous roofing, blistering of the membrane may result. When water vapor enters a building from the ground and subsequently condenses, condensation may manifest itself as peeling paint or rotting roof timbers. Figure 1-2 illustrates loss of adhesion from moisture.

The harmful effects of water on building materials cannot be emphasized enough. The number of harmful effects is endless and includes corrosion, efflorescence, decay, blistering, and dimensional change. If the entrance of water can be controlled through proper detailing, a building can be made more durable, with considerable reduction in maintenance and repair.

Water Absorption

The absorption of water into a building material or system may result in a number of deleterious affects. The following examples should alert the designer and specifier when making materials selections, design decisions and field inspections.

1. *Dimensional Changes.* An increase in moisture content may cause a corresponding dimensional or volume expansion. This phenomenon may occur in brick, mortar, and concrete subjected to rain. As the moisture content decreases, a corresponding reduction in dimension occurs. Thus, changes in moisture content may cause stresses resulting in shrinkage cracks or breaking away from adjacent materials.

2. *Chemical Attack.* The absorption of water into a building material or system may induce a chemical reaction such as corrosion of metals, acid rain attack on metals, stone and concrete and sulfate reaction with concrete. This chemical reaction does not take place in the absence of water. Therefore, careful detailing or selective choice of materials will inhibit or negate this chemical attack.

3. *Efflorescence and Leaching.* Liquid water entering into masonry, concrete, or mortar may dissolve certain soluble salts and cause leaching of these salts to the surface as the water migrates outward. This may result in efflorescence on the exterior face of the material or crystallization behind the surface, which may cause surface rupturing. See Figure 1-3 for effects of efflorescence.

4. *Blistering.* Occasionally when built-up roofing is installed there is a danger of entrapping water in the system as the work progresses. In that event, there is the possibility that the sun's heat will result in a blister formation within the built-up roof system.

5. *Freeze—Thaw.* Temperature alone generally has no serious effect on materials. However, when there is absorption of water into a material the effect of freezing can be destructive, depending on the pore

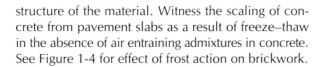

FIGURE 1-3 Efflorescence due to leaching of soluble salts from masonry.

FIGURE 1-4 Effect of freeze–thaw reaction on brickwork. Note spalled rubble at base.

structure of the material. Witness the scaling of concrete from pavement slabs as a result of freeze–thaw in the absence of air entraining admixtures in concrete. See Figure 1-4 for effect of frost action on brickwork.

Permeability–Water Vapor Transmission

Vapor diffusion (water in a gaseous state) is another example whereby water enters into a structure through assemblies and is responsible for deterioration of building materials and assemblies. The maximum amount of water that can exist in the gaseous state is dependent upon the temperature of the air. As the air temperature rises, its ability to hold more moisture increases. When air at a given temperature has absorbed all the vapor it can, the air is saturated or has 100% relative humidity (RH). If the air contains only half as much water vapor at the same temperature, it is said to have 50% RH.

The movement of water vapor is generally independent of air movement. Vapor actually moves by diffusion from areas of high vapor pressure to areas of lower pressures; air with more vapor has a higher vapor pressure.

Condensation is the process in which water vapor changes into a liquid. When warm air laden with moisture is cooled, or touches a cool surface, it reaches its "saturation" temperature and condenses and changes back to water. The temperature at which it occurs is called the *dew point*.

The control of moisture movement in the form of water vapor is essential. A certain amount is necessary for life and comfort. In some industrial buildings low relative humidities in the 15 to 25% range are required. In other situations relative humidities in the 40% range are required. The permissible relative humidity of a given area is the calculated RH which that construction can tolerate, for a given outside temperature and relative humidity without developing condensation on any critical surfaces.

The selection of building materials to reduce the flow of vapor in order to control condensation is therefore critical. The vapor flow resistance of a material is the inverse of the ability of the material to permit vapor to flow, or its *permeance*. The unit of permeance is called the *perm*, which is defined as one grain of water passing through 1 square foot in 1 hour under the action of a vapor pressure differential of 1 inch of mercury, or $P = 1$ grain/h-ft^2 in. Hg.

TABLE 1-3
WATER VAPOR PERMEANCE OF BUILDING MATERIALS

Material	Perm
4-inch brick	0.8
¼-inch plywood	0.72
¾-inch plaster on metal lath	15.0
1 mil aluminum foil	0.0
polyethylene, 4 mil	0.08
polyethylene, 6 mil	0.06
roofing felt, 15 pound asphalt, saturated	5.6
two coats asphaltic paint on plywood	0.4
two coats enamel on gypsum plaster	0.5–1.5

Vapor barriers are materials or systems which retard the flow of vapor under specified conditions. In residential construction a vapor barrier having a perm rating not exceeding 1 is considered adequate. In cold storage buildings a perm rating of 0.01 is considered a high limit. The term *vapor barrier* therefore is not to be construed as a material which offers complete resistance to the flow of vapor. *Vapor stop* would be a more appropriate term in that context.

Table 1-3 illustrates the water vapor permeance of some building materials.

Moisture Expansion and Drying Shrinkage

Some building products that have the capacity to absorb water will expand upon the introduction of water and contract again on drying. Certain porous materials such as masonry, concrete block, concrete, and mortar will behave in this manner, as will wood and fibrous material such as carpets.

Moisture deformation is generally reversible for some materials, such as wood doors which swell with high humidity in summers and shrink again in winter. For other materials such as concrete, mortar, and plaster, the initial shrinkage that occurs during drying and curing may be irreversible.

To offset the effects of expansion and contraction due to moisture gain or loss, some building elements are assembled with a provision for movement. Failure may occur when clearances are insufficient, slotted connections freeze and do not allow movement, or the degree of movement is larger than sealant joints can tolerate.

The specifier and the designer must select materials that will make allowance for movement and detail the assembly to avoid deformation, and cracking failure. In addition, the control of humidity and water during the construction period must be observed and certain operations performed under controlled conditions to avoid prolonged exposure of certain materials to high humidity.

Hygiene, Comfort, Safety

Occupants of buildings may be affected by the selection of products which may have an adverse reaction on them due to a variety of problems associated with health, safety, and/or comfort. With some products, one would have to be a clairvoyant to forecast the dangers inherent in a product on an individual many years later. For example, the problems associated with the use of asbestos were not recognized immediately, since it took 15 to 20 years for the diseases attributable to asbestos exposure to manifest themselves in individuals.

Toxicity

Aside from toxic fumes resulting from combustion, there are building products which may create toxic and noxious fumes as a result of degradation of the chemicals inherent in the product or the release of volatile solvents contained therein.

Recent studies seem to suggest that certain formaldehyde insulation formulations may be emitting gases that cannot be tolerated by some individuals.

Tar products have been known to produce skin irritations and rashes. Roofers tending tar kettles or applying built-up tar roofs have had minor skin cancers that have been traced to tars. Some paint materials using highly volatile solvents must be mixed and applied under controlled conditions, with the applicators wearing respirator devices to safeguard themselves against exposure.

Vermin Infestation

Organic building products are subject to attack by the lower forms of plant life, including fungi, bacteria, and algae. The building products affected are primarily wood, and oils contained in some paint products.

The conditions under which the organisms responsible for this type of degradation grow and multiply best are warm temperatures, water, oxygen, and sometimes light. By introducing mildewcides, bactericides, and other types of inhibitors, manufacturers can reduce the occurrence and degree of vermin infestation in those materials subject to attack under the ideal prevailing conditions.

Slip Resistance (Safety)

A safety factor related to the selection of flooring or paving is slip resisitance. In addition one must consider whether the areas within which the flooring materials will be used will be wetted by water spillage, wet cleaning, rain, or some other source of water.

A measure of the comparative slip resistance of a flooring material under dry or wet conditions can be determined by using several test methods as follows:

ASTM F489   Test for Static Coefficient of Friction of Shoe Sole and Heel Materials as Measured by the James Machine

ASTM F609   Test for Static Slip Resistance of Footwear Sole, Heel, or Related Materials by Horizontal Pull Slipmeter (HPS)

Military Spec. MIL-D-3134H   Deck Covering Material Par. 4.7.6 Non-Slip Properties

The ASTM test procedures noted above can be evaluated by ASTM F695, Standard Practice for Evaluation of Test Data Obtained by Using the HPS Machine or the James Machine.

Mildew Resistance

Under the heading of Durability in this chapter, reference is made to the bactericidal attack of fungi on some building products, causing a degree of deterioration. Not only do fungi, bacteria, and algae effect the durability of building products, but their ability to feed on organic building products may have an effect on the health of the building occupants. It would therefore be prudent to investigate the materials to be selected and provide measures to inhibit growth and propagation of the fungal or bactericidal attack.

Air Infiltration

Comfort for the occupants of a building can be affected by air infiltration, which may produce whistling noises, admit dust, and cause condensation. The extent of air leakage depends upon the design of the building enclosure, the quality of the materials and work, and the action of air pressure differences across openings.

The amount of air infiltration through exterior windows, doors, and curtain walls can be determined by ASTM test method E283. The use of weatherstripping will reduce the amount of air leakage, including the infiltration of dirt and moist air.

DURABILITY

Durability includes the dimensional stability of a material as well as its ability to withstand the rigors of wear, weathering, and other disintegrative influences. An investigation of durability would include the following:

Resistance to Wear

Abrasion

Scratching

Scrubbing

Scuffing

Weathering

Freeze–thaw

Ozone

Fading

Chemical fumes

Bactericidal

Ultraviolet (UV) radiation

Adhesion of Coatings

Delamination

Blistering

Dimensional Stability

Shrinkage

Expansion

Volume change

Mechanical Properties

Resistance to splitting

Resistance to bursting

Resistance to tearing

Resistance to fatigue

The property *durability* assessed in its broadest concept, includes stability against human-made hazards (i.e., wear, abrasion, and chemical attack as in industrial plants) as well as the ability to endure the exterior environment (i.e., rain, frost, sunlight, and heat). This property is often difficult to assess, especially for new products, since short-term test results do not lend themselves to long-term extrapolations. Oftentimes the considered judgment of experts is required, based on knowledge and experience rather than on simple pass-fail tests. However, the user must be made aware of the concept of durability so that certain precautions are exercised in material selections considering the "hazard" to which the material will be exposed.

Resistance to Wear

Fitness for purpose should be the prime consideration. For example, acoustical and insulating products perform such functions based on their cellular, fibrous structure; as a result they are generally fragile. When these products are installed on ceilings as acoustical tile or within assemblies as insulation, their exposure to wear is practically eliminated. However, if they are installed where they are subject to human-made hazards their life span becomes considerably reduced.

Abrasion Resistance

The abrasion resistance of a material is its ability to resist being worn away or to maintain its original appearance when rubbed.

Flooring materials are prime examples of materials in place which are most affected by the hazard of abrasion. Selections of flooring materials will be governed by the type of occupancy. An executive office with negligible traffic can be covered with the most delicate of carpeting, whereas entrance corridors to public school buildings will require rugged, wear-resistant materials to withstand volume traffic of a harsh nature.

An early ASTM test for abrasion resistance of a flooring material was and still is C501, Relative Resistance to Wear of Unglazed Ceramic Tile by the Taber Abraser, designed to determine the abrasion resistance of ceramic tile. In the absence of other standard abrasion tests, ASTM C501 was used for materials other than ceramic tile, such as resilient tile and cementitious and resin matrix terrazzo.

At present there are a number of tests to measure the abrasion resistance of several building materials, as follows:

ASTM C779 Abrasion Resistance of Horizontal Concrete Surfaces

ASTM C944 Abrasion Resistance of Concrete or Mortar Surfaces

ASTM D2394 Simulated Service Testing of Wood Base Finish Flooring

ASTM F510 Resistance to Abrasion of Resilient Floor Coverings

ASTM C241 Abrasion Resistance of Stone Subjected to Foot Traffic

In addition to wear variation of floors one can also measure the wear resistance of paints and coatings by several standard test procedures as follows:

ASTM D968 Abrasion Resistance of Coatings of Paint by the Falling Sand Method

ASTM D821 Evaluating Degree of Abrasion, Erosion in Road Service Tests of Traffic Paint

Scratching, Scrubbing, Scuffing

Materials that may be subjected to scratching should be investigated to determine whether the manufacturer offers suitable touch-up products that can cosmetically hide the scratches or whether other satisfactory procedures can be utilized to eliminate or minimize them.

Materials that will be subjected to scrubbing and/or scuffing require test data to indicate the expected number of scrubbings and/or scuffings before replacement is required. For example, certain scrub tests exist to ascertain the longevity of paints and textiles exposed to repeated scrubbings, and similar test methods exist for floor polishes exposed to markings, scratching, and scuffing.

Weathering

Materials exposed on the exterior of a structure will be subjected to the rigors of the particular environment where the structure is located. Temperature, rainfall, UV, ozone, chemical industrial fumes, fungal attack, will each have a deleterious effect on some materials. Some materials will withstand these elements better than others. Materials indigenous to a specific climate

FIGURE 1-5 Caryatids at Acropolis temporarily protected prior to removal to a museum.

will exhibit minimum deterioration in that locale but will disintegrate over a period of time when used in other climes. Witness the use of brownstone for buildings in New York City. Quarried in the Midwest, the material eroded after years of exposure in the New York environment. The marbles and granites used on the Acropolis in Athens weathered successfully for thousands of years until the exhaust fumes of modern-day industry and automobiles intermixed with water produced an acid rain that weathered the stone appreciably in a matter of years—so much so that the Caryatids of the Erechtheum have been removed to a museum to safeguard them against further deterioration (see Figure 1-5).

It is essential to understand that few materials are in and of themselves durable. Interaction of the environment with the material will determine its durability. Oftentime coatings (organic or metallic) will prolong the life expectancy of a material. Occasionally the detailing of certain features will have an effect on the life span of the materials or assemblies.

Freeze–Thaw

A reference to freeze–thaw was made under the heading of Habitability–Water Permeability. Materials having the capacity to absorb water—especially masonry, cementitious materials, and soils—may experience a freeze–thaw phenemenon which induces stress leading to either minor or major failures, such as cracks or disengagement from the structure depending on the severity of the condition. Air entrainment in concrete has reduced the effects of freeze–thaw. Similarly, frost susceptibility is reduced in soils when fines below sieve size 200 are less than 10%.

Ozone

Ozone is oxygen containing three rather than the normal two atoms of oxygen. Ozone is prevalent in the upper atmosphere and also is produced at ground level by electrical discharges. Since it is unstable, it is extremely reactive. Building materials that oxidize, primarily rubber and certain plastics unless properly compounded, will react with ozone and become brittle or crack. An ASTM test method to determine the degree of deterioration of rubber by ozone is D1149.

Fading (Colorfastness)

Some materials exposed to sunlight can undergo a fading of color. The degree to which material colors are fast or tend to fade can be determined by certain test procedures as follows:

ASTM C798   Color Permanency of Glazed Ceramic Tile

ASTM C538   Color Retention of Red, Orange and Yellow Porcelain Enamel

ASTM D1543 Color Permanence of White Architectural Enamels

ASTM G45   Fading and Discoloration of Non Metallic Materials

Chemical Fumes

Gases that can enter into the weathering process are certain pollutants such as ozone, sulfur dioxide, oxides of nitrogen, and chemical fumes that are the by-products of industrial plants. These gases in combination with rain, sand, dust, and wind can impinge upon exposed surfaces and cause degradation of coatings, organic building materials, metals, and masonry materials.

The chemistry of the building material selected for a given exposure and the nature of the environment may cause a chemical reaction. Thus, it is essential to select a material that will better maintain its stability and durability in a known industrial environment.

Copper gutters and flashings can survive in most environments. However, when these structures are located near coal burning plants, sulfurous fumes will attack and pinhole the metal.

The simple solution of providing a coating that can resist the chemical fumes is all that is required to overcome the problem. Sometimes minor changes in the chemical make-up of a manufactured material can have a profound influence on its ability to resist chemical attack.

Bactericidal

The most prevalent forms of attack on building materials are fungal decay of wood and mildew of certain paints.

Fungi feed on and therefore decompose a wide variety of organic building products, particularly wood and oil-based paints. Paints based on synthetic resins do not support mildew growth. However, mildew can grow on latex paints which have thickeners or emulsifiers that contain cellulose or protein derivatives.

For fungi to grow and prosper, certain conditions must be met. These include moderately warm temperatures, water, and a source of food such as cellulosic products or proteins and fats.

In an environment which is ideal (temperature, water, and food) fungi will thrive. As the environment becomes less ideal, the growth becomes slower.

For wood near soils or water, wood-preservative treatments are usually sufficient to discourage fungi attack.

For paints and coatings, the addition of a mildewcide to the paint will be sufficient to deter mildewing.

Mildew resistance tests for paints are covered in Federal Test Standard No. 141, method 6271.1, and in ASTM D3274, Evaluating Degree of Surface Disfigurement of Paint Films by Fungal Growth.

Ultraviolet (UV) Radiation

Electromagnetic radiation is energy propagated by an electromagnetic field. The electromagnetic spectrum runs the gamut from gamma rays with the shortest wavelengths to microwaves with the longest wavelengths (see Table 1-4).

Ultraviolet radiation lies in the wavelength range of 10 to 380 nm, just below the visible light spectrum.

Radiation, particularly UV, has an effect on the durability of certain building products. It acts by changing the chemical structure of polymers used in organic building products and thereby affecting their physical characteristics. Fortunately the intensity of destructive UV is a small part of the total solar radiation and is reduced as the angle of the sun decreases and by clouds and smoke. However, sufficient UV is received in southern climes to cause degradation of building materials, particularly in the presence of oxygen, water, and heat.

TABLE 1-4
ELECTROMAGNETIC SPECTRUM

Name	Wavelength Range
Gamma rays	0.01 – 1 Å
X-ray	1 – 100 Å
Ultraviolet (UV)	
Extreme UV	10 – 100 nm
Far UV	100 – 200 nm
Middle UV	200 – 300 nm
Near UV	300 – 400 nm
Visible light	400 – 770 nm
Infrared (IR)	
Near IR	770 – 2500 nm
Middle IR	2.5 – 30 μ
Far IR	30 – 300 μ
Hertzian waves	
Microwave	0.3 – 100 mm
Radio and television	0.1 – 1000 m

Natural rubber, neoprene, butyl, EPDM, and some sealants will experience UV degradation unless properly compounded with antioxidants and UV absorbers.

Artificial weathering tests, carbon arc, and mercury lamps have been used to determine durability when exposed to UV. More recently Xenon light weatherometers, ASTM G26, which duplicate natural UV more closely, have been used to test the effect of exposure of building materials to UV.

Adhesion of Coatings

To some extent, under the heading Habitability—Water Permeability, the subject of water infiltration has been discussed with its attendant problems. While paints and coatings are to some degree water repellants and retard the flow of water inward, the vapor-permeable type will permit water vapor to move outward as well under favorable conditions. However, when water enters through failures in the paint film of an impervious coating, blistering may occur as the result of water vapor buildup. This is induced by the effect of the sun causing a heat buildup with subsequent evaporation of the water in the substrate.

Dimensional Stability

As a result of both daily and seasonal changes in air temperature, solar heating, and radiative cooling,

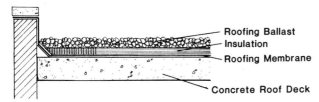

FIGURE 1-6 Inverted insulated protected roof system.

temperature variations occur that cause building materials to undergo dimensional changes including expansion, contraction, and volume changes.

In the design of a building, when one selects materials to be used on the exterior, the properties of the material or of those used in combination must be completely understood, since the exterior environment will have an effect on dimensional stability. It may be necessary to locate the materials where mechanisms that would cause dimensional instability are least active.

For example, built-up roofing will expand and contract when exposed to the elements. There will also be volume changes in the asphalt or tar used as the flood coat. By changing the location of the built-up roofing so that the insulation is installed above the roofing, the dimensional change in the built-up roofing is minimized. Figure 1-6 illustrates this improvement in roofing.

Mechanical Properties

Materials and components that exhibit movement may be subjected to splitting, bursting, tearing, and fatigue. Movement may be induced by temperature change, moisture content changes, and application of loads such as wind, snow, and occupancy. Hence, the durability of materials and components is a function of the stresses developed as a result of temperature and moisture changes and application of loads.

Deformation due to Loading

Stress in a building material may be induced by wind loads, snow loads, or occupancy loads. The amount of stress depends on the magnitude of the load. With loading, an unrestrained material deforms, although the order of magnitude may be small. Many materials have elastic properties and up to certain load limits will resist permanent deformation. The ratio of stress to strain is known as the modulus of elasticity (E) where

$$E = \frac{stress}{strain} \text{ or } \frac{load \text{ per unit area}}{deformation \text{ per unit dimension}}$$

This constant of proportionality, the ratio of stress to strain, represents the inherent ability of the material to resist elastic deformation. The elastic limit is the greatest stress which a material is capable of sustaining without permanent deformation upon complete release of the stress. When the elastic limit is exceeded, a permanent set may be induced in the material. Materials that observe this stress–strain ratio are metals, wood, concrete, and masonry. Organic materials such as built-up roofing, elastomeric roofing, and sealants, and materials such as glass do not observe this elastic ratio because they are plastic materials and also because their mechanical properties change with aging and weathering.

However, even the materials that fall within the parameters of the stress–strain range of deformation will experience another type of degradation when subjected to sustained or long-term loading, which will not fully disappear when the loading is removed. This deformation is called creep or plastic flow.

Temperature Changes

Mechanical properties of materials will change with temperature. Physically, a change in temperature, in addition to change in length, will alter the hardness and strength of a material. Thermoplastics, such as neoprenes, PVCs, EPDMs, whose temperatures may rise with exposure to the sun, will soften and result in a diminution of the tensile strength. In extreme cases this may lead to bursting, tearing, or splitting. Sealants that are properly formulated will resist forces imposed by elevated temperatures. Low temperatures will cause sealant joint widths to expand as the panel sections between them contract. Since the sealants are expanding while they are simultaneously getting colder and stiffer, an added burden is imposed on their ability to perform. The rate of temperature change is an important factor with these materials, since organic compounds can generally tolerate slow rates of strain more readily than fast rates of strain.

Moisture Content Changes

The freezing of water in porous materials such as concrete, stone, brick, and concrete block is in some measure the cause of cracking, splitting, or bursting. Frost failure occurs in materials that are both wet and cold. Materials having high porosity and small pores are generally more susceptable to this phenomenon, and alternating freeze–thaw cycles contribute to splitting and scaling of these materials.

COMPATIBILITY

In the design of buildings there are very many situations where differing materials are joined together as the result of detailing specific configurations. When differing materials are in contact with one another on the exterior of a building, they are subjected to moisture, temperature changes, radiation, and oxygen. These external weathering events can aggravate the incompatibility of the materials, leading to corrosion, degradation, and material failure.

Less frequently, incompatibility of materials may occur as a result of a chemical process or the emanation of fumes within a structure that may have a corrosive or other deteriorating effect.

Compatibility includes the ability of materials and systems to withstand reaction with adjacent materials in terms of chemical interaction, galvanic action, and degradation of physical properties.

The most common experience with incompatibility of exterior materials is that of bringing dissimilar metals in contact with one another. In the presence of moisture a galvanic cell is created, causing a flow of current from one metal (the anode) to another metal (the cathode), resulting in the corrosion of the metal serving as the anode at the expense of the metal serving as the cathode. For example, steel nails used in conjunction with copper flashing or roofing will act as anodes and corrode.

This phenomenon of galvanic action is referred to as the electromotive force (emf), which is an electrical potential difference that causes the movement of electricity or tends to produce an electric current. The electromotive force series (EMF Series) is a listing of chemical elements arranged according to their standard electrode potentials. The "noble" metals such as gold are positive, the "active" metals such as zinc are negative. The "noble" metals at the bottom of the EMF series are the least susceptible to corrosion; the "active" metals at the top of the Series are the most susceptible. The EMF Series for metals used in building construction is shown in Table 1-5.

To eliminate the possibility of galvanic action, felts, paint coatings, or bitumen have been employed to separate dissimilar roofing metals. Conversely, this principle of EMF has been used advantageously to prevent corrosion by creating a slight positive charge

TABLE 1-5
ELECTROMOTIVE FORCE (EMF) SERIES

Magnesium
Zinc
Aluminum
Chromium
Iron or steel
Stainless steel
Tin
Lead
Copper
Brasses
Monel
Silver
Gold

at structures such as water towers and underground pipe lines, to reduce galvanic electrical conductance and thereby eliminate corrosion.

The problems of incompatibility were dramatically increased with the advent of new human-made chemical building materials after World War II. A case in point was the introduction of modern-day sealants. An early sealant was utilized for the sealing of joints in marble on the United Nations Building in New York. Within a short period of time, a pinking discoloration was noticed on the white marble that was ultimately attributed to the sealant. By reformulating the sealant, the discoloration of the marble was overcome.

Endless tales of similar mishaps have occurred when dissimilar materials brought together for the first time, exhibited either chemical or galvanic reactions. To reduce the possibility of incompatibility, it is essential to require that manufacturers of materials be apprised of the proposed use of specific materials within a certain detail and configuration and perform necessary tests to determine possible incompatibility prior to use.

Incompatibility also includes the differential thermal movement. Again, when diverse materials are used within the same configuration, their rates of thermal expansion should be checked to make certain that they are not so disparate as to cause buckling or separation because of uneven thermal expansion or contraction.

When paints or coatings are resurfaced with different coatings, tests should be performed to ascertain the compatibility of the new coating systems to prevent lifting, softening, emulsification, and degradation that may result from chemical incompatibility of the new and the old systems.

SITEWORK

Architectural materials utilized for sitework are associated with designs for the paving and surfacing of such areas as walks, plazas, platforms, steps, roads, and promenades.

Engineering materials usually employed for streets, roads, and walks are asphaltic concrete and portland cement concrete. These are generally designed by engineering specialists, except that color, texture, and patterns are generally controlled by architects. Even the so-called blacktop (asphaltic concrete) can assume a grayish rather than a black appearance by the introduction of broken stone of a grayish hue seeded into the surface. The architect will also lay out patterns in concrete paving and make decisions on texturing the concrete to likewise control the architectural aesthetic qualities of walks and paved surfaces.

Architectural materials used for paved surfaces are myriad and include brick, clay tile, terrazzo, stone, precast concrete blocks, and quarry tile. In many instances information on these materials will be found under separate chapter headings. This chapter discusses similar materials unique to sitework and methods of setting.

PAVING

BRICK PAVING

Material

Paving brick is a vitrified, dense, hard brick made from fire clay or shale designed for use where abrasion and wear are important factors. ASTM C902 Specification for Pedestrian and Light Traffic Paving Brick is a standard for paving brick. Paving brick is available in thicknesses as thin as ½ inch and up to 2 ½ inches in thickness and in various face dimensions. In addition to paving brick, both building brick and facing brick are utilized for brick paving where certain colors and textures are unique to specific projects.

Setting Brick Paving

Brick intended for walks, pavements, plazas, and so forth may be set in mortar, on a properly prepared sand bed or on bituminous asphalt setting beds. The latter utilizes a mixture containing 7% asphalt and 93% fine aggregate. Patterns may include herringbone, running bond, and so forth.

PRECAST CONCRETE PAVING

A number of manufacturers have developed precast concrete paving block units (pavers) of unusual shape and in a variety of colors. The pavers have compressive strengths of about 8500 psi. The concrete pavers are generally laid on a dry sand bed with tight joints, and the joints are filled by sweeping sand into them.

STONE PAVING

In Chapter 4, under heading of Building Stones, a number of stone materials are described. All of them may be used for paving in limited degrees. In addition to those building stones, several others such as bluestone and quartzite are also used. Paving stones may consist of large slabs laid in an ashlar pattern, or small, irregularly shaped stones laid out in a random pattern.

Flagstones are thin slabs of stone used for flagging or paving. They are generally fine-grained sandstone, bluestone, quartzite, or slate, but thin slabs of other stone may be used for this purpose.

Generally, paving stone may be as thin as ⅞ inch. However, the thickness should be a function of the face size to prevent breaking during transportation and handling. Stone producers should be consulted to ascertain minimum thickness requirements relative to face size selections.

Abrasion Resistance

Since wearing by abrasion is an important physical characteristic, a test method to determine abrasion resistance is an important guide to the selection of stone. ASTM C241, Test for Abrasion Resistance of Stone Subjected to Foot Traffic is such a useful tool. Generally, the granites have the highest degree of resistance to abrasion. A ranking of abrasion resistance as conducted by the Bureau of Standards is as follows:

Material	Abrasion Resistance
Granite	37 to 88
Limestone	1 to 24
Sandstone	2 to 26
Slate	6 to 12
Travertine	1 to 16

Slip Resistance

When selecting stone surfaces that may be wetted from time to time due to exposure to the weather, consideration must be given to the type of finish. Obviously, a polished finish would be ruled out immediately and perhaps honed finishes for some stone species.

The rough-sawn or shot finishes and the natural

cleft finishes are ideal where antislip qualities are desirable as is the thermal finish.

To determine antislip, resistance, laboratory tests using the James Machine or the horizontal pull slipmeter (HPS) described in Chapter I under Durability—Hygiene, Comfort and Safety, may be utilized.

See Chapter 4, for finishes of granite, marble, travertine, sandstone, slate and limestone.

Setting Beds

Stone paving is generally set in mortar, although some may be set in sand beds. Modified latex mortars that can reduce the thickness of setting beds are also used frequently for stone setting. When granite, marble, limestone and travertine are set in mortar, they should be set in nonstaining waterproof cement mortar.

Mortar is generally composed of one part portland cement to three parts sand. Nonstaining cement should comply with ASTM C91. Stones should be placed on setting bed mortar and tamped with a rubber mallet until firmly bedded and then removed. The back of the stone should then be parged with a grout of wet cement, and returned to its initial position and tamped into place.

When set on sand, the beds should be firmly tamped to preclude settlement.

Materials

Bluestone

Bluestone is a dense, hard, fine-grained sandstone of bluish gray color that splits easily along original bedding planes to form thin slabs. It is also available in buff, lilac, and rust colors.

Finishes for bluestone include:

Natural Cleft: Having the natural seam split.
Rubbed: Rubbed with coarse industrial diamonds after wire sawing.
Thermal: A flame-textured finish obtained by heating.

Exposed edges of bluestone may be diamond sawed, rubbed, snapped, rocked, or thermal. Bluestone has a wear and abrasion resistance of Ha-49 when tested in accordance with ASTM C241.

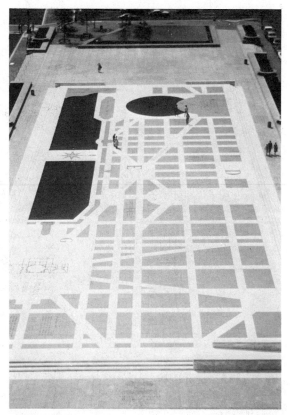

FIGURE 2-1   Pennsylvania Avenue. Development, Washington, DC: granite—Sunset Red, Academy, Carnelian, and Bright Red; thermal finish. (George E. Patton/Venturi & Rausch, Architects.) (Courtesy of Cold Spring Granite Company.) This illustration also appears in the color section. This illustration also appears in the color section.

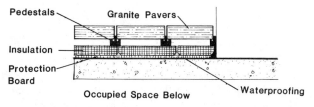

FIGURE 2-2   Granite pavers mounted on pedestals.

Granite

Granite* paving may consist of large panels of stone for pedestrian use or small blocks intended for vehicular use or pedestrian malls or walks (see Figure 2-1). Granite is particularly suitable as a paving material when mounted on pedestals for plaza areas, over occupied areas, as shown in Figure 2-2.

*See Chapter 4 for physical characteristics, colors, and finishes.

FIGURE 2-3 American Republic Building, Des Moines, IA: gray marble flooring on interior carried into outdoor courts. Skidmore, Owings & Merrill, Architects (Photographer, Ezra Stoller © ESTO.)

FIGURE 2-4 LBJ Library, Austin, TX: travertine paving. Skidmore, Owings & Merrill, Architects (Photographer, Ezra Stoller © ESTO.)

Limestone

Limestone* for use as paving in commercial and public areas should have a minimum abrasive hardness of 10 as per ASTM C241. It is also essential to provide proper slope to drain the surface to avoid moisture collection or ponding. If limestone is contemplated for exterior paving or steps, it would be prudent to consult with the supplier or producer to obtain information on details and limitations of the material.

Marble

Only Grade A marble* should be considered for exterior paving. Thickness is a function of face size with respect to handling, transportation, and setting. The marble supplier and quarry should be consulted for optimum thickness requirements. See Figure 2-3 for typical marble installation.

*See Chapter 4 for physical characteristics, colors, and finishes.

Sandstone

The sandstones* exhibit excellent wear and abrasion-resistant characteristics and make good exterior paving materials. They are more typical of installations known as flagging where utilitarian rather than aesthetic qualities are paramount.

Slate

When slate* is used, edges for joints in the field of paving are generally sawn. Where edges are exposed at platform ends or stair treads, they may be rubbed or honed. A minimum abrasive hardness of 8 as per ASTM C241 should be specified.

Travertine

Crevices, craters and holes in travertine* used for paving may be filled with travertine chips and mortar

FIGURE 2-5   U.S. Steel—One Library Plaza, New York, NY: rustic terrazzo paving. Skidmore, Owings & Merrill, Architects. (Photographer, Ezra Stoller © ESTO.)

rubbed finish is best used for exterior surfacing. See Figure 2-4 for exterior travertine paving.

TERRAZZO

For exterior terrazzo paving see Chapter 9, rustic terrazzo. Figure 2-5 illustrates a structural rustic terrazzo installation.

RETAINING WALLS, GARDEN WALLS, AND SOLAR SCREENS

Architectural materials employed for sitework in connection with walls and screens of various types include architectural concrete, brick, block, stone, wood, and metal. For the most part these materials and their applications for usage are contained under other chapter headings.

3

ARCHITECTURAL CONCRETE

With the invention of portland cement in 1824 by Aspdin, an Englishman, and its subsequent manufacture in the United States in 1871, concrete became a basic building material for use as a structural element in foundations, framing, and floor systems.

Since concrete is a plastic, moldable material, architectural designers have long attempted to experiment with concrete as an exposed architectural expression. An early example of architectural concrete is Frank Lloyd Wright's First Unity Temple in Oak Park, Illinois, built in 1906 (see Figure 3-1).

It was not until after World War II that sufficient technical know-how was accumulated by architects and engineers and painstaking reeducation of concrete contractors was undertaken to achieve and construct credible designs in both cast-in-place architectural concrete and architectural precast concrete.

FIGURE 3-1 Architectural concrete as utilized by Frank Lloyd Wright, First Unity Temple, Oak Park, IL. (Photographer, Marvin Tetenbaum)

CONCRETE USAGES

The use of concrete may be channeled into a number of areas as follows:

Structural
Pile foundations
Footings
Foundation walls
Structural members—Floors, walls, roofs

Special Techniques
Slip form
Lift slab
Tilt-up
Prepacked concrete

Architectural Forms
Ribbed slab
Waffle slab
Flat slab
Thin shell
Folded plate

Paving
Highways
Air fields
Roads
Walks
Ramps and steps

CONCRETE INGREDIENTS

TABLE 3-1
TYPES OF PORTLAND CEMENT

Type	Use
I	Standard portland cement for general use when special properties are not required.
II	For use where moderate sulfate resistance or moderate heat of hydration is desired.
III	For high early strength, developing almost twice the strength of Type I at three days. Used especially for cold weather concreting.
IV	For use when a low heat of hydration is desired, as in mass concrete for dams to diminish cracking or in other massive members. It is also slow setting and attains its strength over a longer period.
V	A sulfate-resisting cement for use where severe sulfate conditions are encountered and where alkaline waters or soils are present.

Concrete is a mixture of portland cement, aggregates, and water. In addition, admixtures are often incorporated to impart certain characteristics, and reinforcment is used to increase tensile strength.

PORTLAND CEMENT

ASTM C150, the standard specification for cement used in the United States, lists five types of portland cement, as shown in Table 3-1.

CEMENT COLORS

Standard portland cement is usually a grayish color. There are also some manufacturers that have buff-colored cement and others that manufacture a white cement. Both buff and white cement are made to ASTM C150 Standards. White cement may be used also as a base for the addition of pigments to obtain a variety of colors.

AGGREGATES

Aggregates are inert materials that do not react with cement and water. They generally consist of gravel, crushed gravel, broken stone, and sand. Aggregates are defined by size as coarse and fine, with coarse aggregate being that fraction which is retained on a ¼ inch sieve and fine aggregate being that which passes a ¼ inch sieve. Both coarse and fine aggregate are specified by ASTM C33, which sets standards for grading, deleterious substances, and soundness. In special cases where desired by the architect, fine aggregate may be produced by crushing the specific coarse aggregate chosen.

Gap-Graded Aggregates

Aggregates are graded so that the finer particles fill the voids between the larger aggregates. This reduces the amount of cement paste (cement and water) required and results in a more economical mix. However, for some architectural applications, gap-graded aggregate produces a very pleasing appearance. In a normal gradation of coarse and fine aggregates, the end result of an exposed aggregate finish is a somewhat nonuniform distribution of the aggregates. In a gap-graded mix there is a large percentage of coarse aggregate and a small percentage of fine aggregate, with no aggregate in the intermediate size range. As a result, gap-graded mixes show a uniform size distribution of the exposed aggregate. See Figure 3-2 for an example of a gap-graded, cast-in-place concrete.

Most normally graded concretes have matrix volumes (air, water, cement, and sand) of 55% or more. Gap-graded concrete reduces this matrix to 45–50%, which accounts for the larger proportion of visible coarse aggregates.

A laboratory test program initiated by the Portland Cement Association and described in their *Development Department Bulletin D90* establishes criteria for design mixes, slump, and air content. These criteria, along with the special care involved in taping form joints and vibration control, provide the technical requirements needed by architects, engineers and contractors to obtain architectually exposed gap graded concrete.

FIGURE 3-2 American Republic Building, Des Moines, IA: Sandblasted, gap-graded, gray granite aggregate. Skidmore, Owings & Merrill, Architects. (Photographer, Ezra Stoller © ESTO.)

Special Aggregates

While gravel and stone are the principal coarse aggregates used in concrete, architects are increasingly searching for decorative aggregates to enhance their architectural concrete designs. Such decorative aggregates include granite, quartz, quartzite, crystalline aggregates, onyx, pebbles, marble, glass, ceramic, alundum and emery. They can be obtained from numerous aggregate suppliers nationwide. See Figure 3-3 for photographs of panels of special aggregates. Also see Figure 3-4 for a structure utilizing a granite aggregate.

When using an aggregate for which there is very little technical information as to its reaction with the cement paste, it is essential to perform a petrographic analysis in accordance with ASTM C295. This analysis will determine the physical and chemical properties of the aggregate and its prospective performance as a concrete ingredient.

Lightweight Aggregate

An aggregate for use in lightweight structural concrete is produced by heating such materials as clay, shale, slate, or blast furnace slag. During the calcining or sintering process, the gases formed inside the material expand it to a lightweight, porous form. The concrete resulting from the use of lightweight aggregates weighs approximately 75 to 110 pounds per cubic foot. Lightweight aggregate is specified to comply with ASTM C330.

WATER

Generally, potable water is satisfactory for making concrete or the cement paste that binds the aggregates together. If the source of water is suspect, it should be analyzed to ensure its use as one of the ingredients of concrete.

When water and portland cement are mixed, a chemical reaction, hydration, takes place, producing a cementitious product (cement paste) that binds the aggregates together and forms concrete.

ADMIXTURES

A concrete admixture is defined in ASTM C125 as ''a material other than water, aggregates, and portland cement used as an ingredient of concrete and added to the batch immediately before or during its mixing.''

If properly designed, proportioned, and handled, concrete should not require the use of an admixture. The proper design mix and the selection of suitable materials should result in workable, durable, watertight, and finishable concrete. However, there are instances when an admixture is desirable—for example, when special properties are required that cannot be obtained by normal methods or as economically. Certain special properties that are primarily of interest to the architect include the following: (1) retarding or accelerating the time of set; (2) accelerating of early strength; (3) increase in durability to exposure to elements; (4) reduction in permeability to liquids; (5) improvement of workability; (6) bonding of gypsum and portland cement plaster; (7) antibacterial properties of cement; and (9) coloring of concrete.

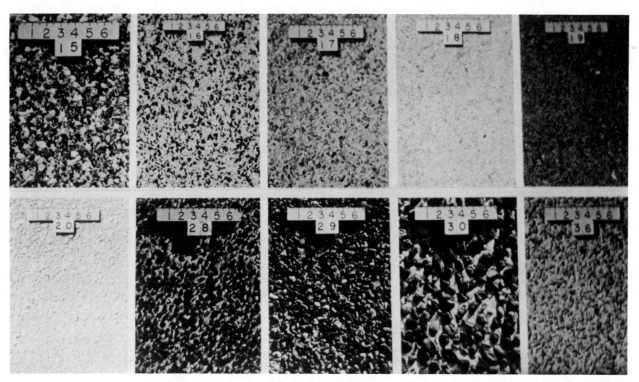

Panel No. 15. Use of three aggregates—verde antique, white marble, and pink granite, on a white background produces a colorful surface.

Panel No. 16. The combination of black obsidian and milky quartz results in a striking contrast.

Panel No. 17. Rose quartz presents a pleasing variety of pink shades against a white matrix.

Panel No. 18. The green of the exposed cordierite aggregate produces a lighter hued surface when viewed from a distance.

Panel No. 19. Pink feldspar is a platy rock that orients itself in the horizontal position during consolidation. The resulting exposed-aggregate surface is very dense and uniform. The large flat crystals reflect light from myriad surfaces.

Panel No. 20. The exposure of very white silica stone on an integrally colored light blue background is most interesting. The distant viewer sees a moderate blue surface with the exposed aggregate barely visible. A closer view shows the mild contrast caused by exposure of the aggregate.

Panel No. 28 Reddish brown Eau Claire gravel contrasts with uniform white background.

Panel No. 29. Use of expanded slag fines results in a coarse-textured white background for the reddish brown. Eau Claire aggregate.

Panel No. 30. A chemical surface retarder produced the correct degree of exposure for the ¾-in. (19-mm) to 1-in. (25-mm) Eau Claire aggregate.

Panel No. 36. Exposure of pink marble on white background provides a subtle coloration.

FIGURE 3-3 Panels of various exposed aggregates. (Used with permission of Portland Cement Association). These photographs also appear in the color section.

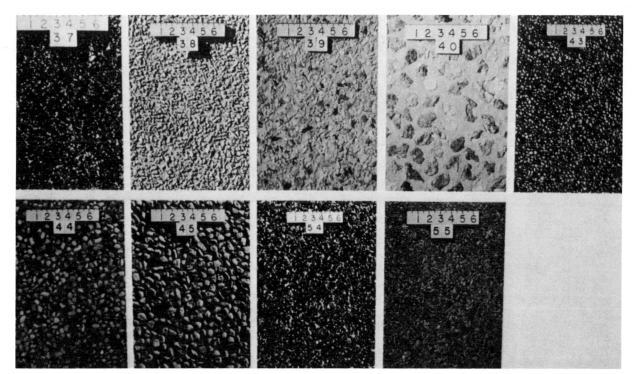

Panel No. 37. Black labradorite rock fractures along crystal faces, and exposed portions sparkle in the sunlight. Aggregate of the size shown here results in a very black surface.

Panel No. 38. White marble in a matrix of silica sand and white cement produces an all-white panel.

Panel No. 39. The brown-yellow-white quartz fractures into flat pieces that orient themselves against the bottom surface of the form. The result is excellent exposure of the quartz.

Panel No. 40. The form was covered with a surface retarder and pink marble was handplaced uniformly on it. White cement-sand mortar was placed over the marble and consolidated. Backup concrete was then placed and consolidated.

Panel No. 43. Dark expanded shale fines produce a dark background that contrasts with the buff-colored Elgin pea gravel.

Panels No. 43, 44, and 45 demonstrate the effect of increasing the size of coarse aggregate. A range of surface retarders produced the correct reveal for the aggregates.

Panel No. 54. A single-colored green glass was used for the exposed aggregate, yet a two-color effect was obtained due to the difference in size of the translucent aggregates. A special glass that was nonreactive with portlant cement was used.

Panel No. 55. Combination of a reflective orange glass with a nonreflective brown natural aggregate makes a very interesting panel. Panels of various exposed aggregates. (Used with permission of Portland Cement Associations.)

FIGURE 3-3 (continued)

Certain special properties that can be controlled or achieved through the use of admixtures, and which are of interest to the structural engineer, include the following: (1) control of alkali-aggregate expansion; (2) reduction of heat of hydration; (3) modification in rate of bleeding: (4) decrease in capillary flow of water; and (5) reduction of segregation in grout mixtures.

It should be stated unequivocally that admixtures are not a substitute for good concreting practices. Admixtures affect more than one property of concrete, sometimes affecting desirable properties adversely. Therefore, trial mixes should be made that reproduce job conditions, using the admixtures with the design mix. The compatibility of the admixture with the other concrete ingredients should then be observed, taking into account its effect on the properties of the fresh and hardened concrete. When special properties are

FIGURE 3-4 Hirshhorn Museum, Washington, DC: Sand-blasted Swenson Pink granite aggregate. Skidmore, Owings & Merrill, Architects. This illustration also appears in the color section.

required, and the use of an admixture is decided on, then those admixtures being considered for use should conform with applicable ASTM Specifications.

The classification of admixtures can be established rather broadly—insofar as architectural specifications are concerned—as follows: (1) air-entraining admixtures; (2) set-controlling (retarding) admixtures; (3) accelerating admixtures; (4) workability agents; (5) dampproofing and permeability-reducing admixtures; (6) bonding agents; (7) fungicidal, germicidal, and insecticidal admixtures; (8) coloring agents; and (9) superplasticizers.

Air-Entraining Admixtures

The admixture that has obtained the widest recognition and use to improve durability of concrete exposed to a combination of moisture and cycles of freezing and thawing, is an air-entrainment agent. The mechanism through which air-entrained concrete resists the disruptive effects of frost action is in the large number of minute air bubbles that are distributed uniformly throughout the cement paste.

Entrainment of the air may be produced by means of air-entraining admixtures added to the concrete in-

gredients before or during the mixing of concrete, or through the use of an air-entraining portland cement. The materials used for air entrainment include natural wood resins, fats, and oils that have been chemically processed. Air entrainment admixtures should meet the requirements of ASTM C260. Conformance with these specifications will ensure that the admixture functions as an air-entraining agent, that it can effect a substantial improvement in the resistance of concrete to freezing and thawing, and that no essential property of the concrete is seriously impaired.

The use of air-entrained concrete has become so widely accepted that cements containing air-entraining admixtures are included in ASTM and Federal Specifications. The admixture is interground with portland cement during the manufacture of air-entraining additions; air-entraining cements should conform to the requirements of ASTM Specifications C226 and C175, respectively.

Set-Controlling Admixtures

Admixtures that delay the setting time of concrete are termed retarders. They are used principally to overcome the accelerating effect of high temperatures during the summer and to delay the initial set of concrete when difficult or unusual conditions of placement are required.

In exposed architectural concrete, set retarders can be used to keep concrete plastic for a sufficiently long period of time so that succeeding lifts can be placed without developing cold joints. Retarders are also used to expose the aggregate in the surface of concrete. This can be achieved by applying a retarder to the forms or to the surface of horizontal planes, thereby inhibiting the setting of the surface layer of the mortar. Upon removal of the forms, the surface mortar is removed by wire brushing or sand blasting, thus exposing the aggregate to produce unusual surface texture effects.

Retarders generally used as admixtures are lignosulfonic acid and its salts, and hydroxylated carboxylic acids and their salts. These should meet the applicable requirements of ASTM C494.

Accelerating Admixtures

Accelerating admixtures are used to achieve high early strength and to shorten the time of set. High early strength results in earlier removal of forms, reduction of required time for curing and protection, earlier use of a structure, and partial compensation for the retarding effect of cold weather.

Chemicals used as accelerators are organic compounds of triethanolamine and calcium chloride. Accelerators should conform to ASTM C494, and calcium chloride should conform to ASTM D98. However, calcium chloride may cause discoloration and its use should be avoided for architectural concrete.

Calcium chloride should not be used in the following situations: (1) in prestressed concrete, because it may cause corrosion of the steel; (2) where aluminum and steel are embedded in concrete because electrolytic corrosion will take place in a humid environment; and (3) in lightweight insulating concrete on metal decks.

Workability Agents

Workability, or the ease with which concrete can be placed, is more often desired by the contractor than by the architect or engineer. Fresh concrete is some-times harsh, and improved workability may be desired for trowel finishing, for placing in heavily reinforced sections, or for placing by pumping or tremie methods.

One of the better workability agents is an air-entraining admixture. The minute air bubbles act as a lubricant and are especially effective in improving workability.

Other workability agents are mineral powders such as bentonite, clay, diatomaceous earth, fly ash, fine silica, or talc. Fly ash and natural pozzolans used as workability agents should conform to ASTM C618.

Dampproofing and Permeability-Reducing Admixtures

The terms "dampproofing" and "waterproofing" imply prevention of water penetration of dry concrete or stoppage of transmission of water through unsaturated concrete. However, admixtures have not been found to produce such results. The terms, therefore, have come to mean a reduction in the rate of penetration of water into dry concrete or in the rate of transmission of water through unsaturated concrete from the damp side to the dry side.

Admixtures for dampproofing include soaps, butyl stearate, and certain petroleum products.

Experts put little credence in the effect of admixtures on the reduction of permeability. The watertightness of concrete depends primarily upon obtaining a well-cured paste having a water–cement ratio not over 0.6 by weight (6¾ gallons of water per bag). Concrete made with less than 5½ gallons of water per bag and well cured, produces a good, watertight concrete that is not improved with the use of dampproofing agents.

Bonding Agents

These admixtures are used to increase the bond strength between new and old concrete, and for bonding gypsum and portland cement plaster to concrete. The admixtures are polymers of polyvinyl chloride, polyvinyl acetates, and acrylics. There are two categories of bonding admixtures: the reemulsifiable type and the nonreemulsifiable types. The latter are water-resistant and therefore better suited for exterior application or in areas where moisture is prevalent.

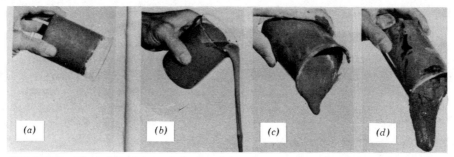

FIGURE 3-5 Effect of plasticizers on the flowability of cement paste. (Reproduced from Canadian Building Digest No. CBD 203.) (A) No plasticizer. (B) with 0.3 percent super plasticizer. (C) with 0.3% lignosulfonate. (D) with 0.4% lignosulfonate.

Fungicidal, Germicidal, and Insecticidal Agents

Antibacterial cements are usually those having an admixture ground into the cement to impart fungicidal, germicidal, or insecticidal properties to the cement. These materials include phenols, dieldrin, and copper compounds, which are useful in tile joints in such areas as locker rooms, food plants, and dairy plants.

Coloring Agents

Pigments added to concrete to produce color are termed coloring admixtures. They should be colorfast, chemically stable, and have no adverse effect on the concrete. These pigments are generally inorganic oxides of the synthetic type.

Superplasticizers

A superplasticizer is a water-reducing admixture that is used to lower the amount of water used in a concrete mix. A normal water-reducing admixture is capable of reducing the water content by about 10 to 15%. A superplasticizer can reduce the water content by about 30%. Superplasticizers are also known as superfluidizers, superfluidifiers, super water-reducers, or high-range water-reducers. They were first introduced in Japan in about 1965. Their basic advantages include (1) better workability of concrete, resulting in easy placement without reduction in cement content or strength; (2) high-strength concrete with normal workability but lower water content; and (3) a con-

crete mix with less cement but normal strength and workability.

The major superplasticizers are sulfonated melamine-formaldehydes (SMF) and sulfonated naphthaleneformaldehydes (SNF), and are available as solids or liquids. Superplasticizers permit excellent workability in that they permit easy and quick placement, especially around heavy reinforcement. Slumps of as much as 8 inches are obtained within a few minutes after the addition of the admixture so that the concrete flows easily and is practically self leveling. Such concrete has the advantage however of remaining cohesive without bleeding, segregation, or loss of strength. See Figure 3-5 for effect of plasticizers on the flowability of cement paste.

Since superplasticizers make possible a high degree of workability, the creation of architectural shapes becomes less restrictive and allows architects the latitude to experiment and develop new forms. Contractors likewise will find that placing of concrete in intricate reinforcing becomes less onerous.

ARCHITECTURAL CAST-IN-PLACE CONCRETE

Since concrete is a free-forming plastic material, architects have long had a fascination with the architectural possibilities that a plastic yet structural material like concrete has to offer. The initial drawback to exploring these possibilities stemmed from the fact that concrete was associated with its long-time role as a structural material hidden in foundations, and covered when used in columns and supports. Additionally, since concrete was not construed as an architectural element, concrete contractors made no effort to develop techniques to safeguard finished work during the construction process. It was not until many years later that enterprising architects working with cooperating contractors developed the necessary criteria to extricate concrete from its use in the foundations of buildings and use it in various forms and finishes as a structural finish material featuring the aesthetic qualities inherent in most finish materials.

Architectural cast-in-place concrete can be achieved quite readily today with very little more of a premium as a result of the pioneering effort of a number of architects and contractors. To ensure a more complete success, it is essential that drawings define the scope of architectural cast-in-place concrete to differentiate it from the normal structural concrete; likewise, there should be a separate specification section identified as "Architectural Cast-in-Place Concrete" so that contractors are forewarned regarding what is expected of them when performing this work. This strategy will alert the contractor to the fact that there are significant differences between this phase of the concrete work and that shown and specified for foundations, slabs, and concealed concrete.

DESIGNING

The decision to use architectural cast-in-place concrete requires an assessment of a number of factors that have an influence on its successful achievement. Proper configuration and size of a member must be considered since form sizes, reinforcing, placing, joint location, and form stripping will affect the member configuration.

The design will include location of vertical and horizontal expansion joints, construction and contraction joints, so that these occur at inconspicuous places or are highlighted to accent certain features.

Architects should avoid the use of large, flat, smooth, uninterrupted expanses of concrete surfaces, since these are the most difficult finishes to achieve uniformly. The use of textured form surfaces, induced textured finishes, and other relief features will minimize minor blemishes, whereas such blemishes will be exaggerated on wide, uninterrupted, smooth surfaces.

Since the structural aspects will have a significant influence on the ultimate outcome, it is essential that the structural engineer be involved at the outset. The engineer's input will be necessary to guide the architect so that the architectural appearance will not be marred by loading and stresses that may induce excessive cracking, spalling, or defects and thus detract from the aesthetic qualities.

PRELIMINARY INVESTIGATIONS

The sources of the major components of concrete—that is, portland cement and aggregates—should be investigated to determine whether satisfactory materials are available locally or whether, in the case of specific unusual aggregates, they must be obtained from distant sources, which will affect the cost.

Samples representing the desired surface, color, texture, and aggregate configuration should be cast, and the selected final architect's sample should be available for inspection and examination by prospective bidders. A minimum size of 18 or 24 inches square by 2 inches thick should suffice to display visual surface characteristics and depth to allow for mechanical texturing.

PRECONSTRUCTION CONFERENCES AND MOCK-UPS

A preconstruction conference should be held at which the architect, engineer, general contractor, and concrete subcontractor are present. At this conference the special requirements of the specifications and drawings can be explained by the architect and engineer, and the contractors can raise any questions regarding aspects of these documents that might impose impediments. Such a conference usually provides the basis for continuing dialogue, cooperation, and coordination throughout the concreting process.

A preconstruction mock-up is a valuable tool in the overall achievement of a successful concreting operation. A full-scale mock-up should be representative of the typical building module and should incorporate all of the elements that would normally be encountered during the course of construction. The building team that will be responsible for its execution should witness the construction of the forms, the placing of reinforcement, the taping of form joints, the placement of concrete accessories, the placement and vibration of concrete, the curing process, the removal of forms, the finishing techniques, and the patching and repairing of surface defects. The mock-up becomes in effect the proving ground for the efficacy of the design, the materials selections, and the construction procedures.

CONCRETE INGREDIENTS

Cement

To minimize color variation, cement of the same type and brand, from the same mill and raw material, should be used for all the architectural concrete on the same project. For large projects, there should be assurance from the manufacturer that ample supplies are available to satisfy the requirements of the project.

White cement can vary slightly in color between brands or mill sources and the same caveats apply to its use as for the normal gray cement. The use of mineral pigments to provide good color intensity and uniformity with white cement should be carefully selected and controlled.

Aggregates

As with cement, aggregates should be specified to come from one source for the duration of the project, and the supply should be adequate to fulfill the project requirements.

Where the surface will not be exposed to reveal the aggregate, the selection of the coarse aggregate will not affect or influence the appearance of the architectural concrete. The fine aggregate in an as-cast finish will, however, influence the appearance, especially if a darker fine aggregate is used with light-colored or white cement.

In cases where the aggregate is exposed, special attention must be paid to its selection and use. Aggregates with proven service records or those subjected to satisfactory laboratory testing and in some cases to petrographic analysis (ASTM C295) will serve admirably. Soft, nondurable aggregates including some marbles, limestones and high-calcium materials are not suitable for exterior use. Aggregates containing iron-based minerals, which react with moisture and cause staining should be avoided for exterior use.

Where the surface of the concrete will be treated after hardening, consideration should be given to the choice of aggregate and the method of treatment (e.g., abrasive blasting, bushhammering, tooling, chemical retardation). For some aggregates the treatment selected may cause the aggregate to become dull or hazy and for others the appearance may be heightened due to the treatment. An early research program and investigation is essential to establish ultimate choices.

Admixtures

Air-entraining admixtures are useful both in increasing durability and improving workability, especially for harsh mixes.

The new superplasticizers offer a multitude of new advantages, that is, higher strength, lower water–cement ratios, and improved workability. However, sufficient data on long-term effects are yet to be accumulated and analyzed. The use of superplasticizers in exposed aggregate surfaces must also be reviewed, since their fluidity may affect uniform appearance.

The use of calcium chloride as an accelerating admixture is not recommended, since it contributes to the corrosion of metal reinforcement and accessories and may mottle or darken architectural concrete.

It would be prudent to restrict the use of admixtures and to obtain preliminary mix design information on those mixes containing admixtures, especially to verify color variation.

DESIGN MIXES

For as-cast finishes, design mixes do not differ materially from those used in good structural concrete practice. Improved workability, however, is essential in order that the concrete be brought into intimate contact with the forms, especially where textured forms are used so that the face pattern of the concrete will be a reproduction of the form face.

Where exposed aggregate finishes are required, the mix design is a factor of the aggregate sizes, depth of exposure, the cement factor, aggregate shapes, slump, and workability. Engineers and concrete testing laboratories qualified in this type of work should be sought out for their expertise.

REINFORCING AND ACCESSORIES

Reinforcing and metal accessories should be located at least 2 inches from an exposed exterior surface to ensure proper coverage and avoid surface staining due to rust. Where metal is closer than 2 inches from the surface, galvanizing should be considered.

Stainless steel tie wire should be used for tying reinforcement, to avoid staining exposed surfaces. The wire should be bent back, away from forms.

Supporting chairs, spacers, or bolsters should be stainless steel or high-density plastic of a color that matches the concrete, or galvanized steel with mushroom plastic-tipped legs of a color that matches the concrete. See Figure 3-6 for concrete accessories.

FORM MATERIALS

Architectural concrete can be textured by one of two basic methods: (1) molding it in the plastic state (off-the-form concrete); or (2) treating the hardened surface. These two methods offer a number of variations that give the architect a wide range of choices. In making decisions concerning selection of forming materials and the economics involved with varying types, it is essential to understand that lumber and plywood (except those with plastic overlays) are absorbent. Absorbent forms can affect the color of the concrete through variations in absorptivity. Increased absorption will reduce the water–cement ratio, resulting in a darkening of the concrete surface. There can be variations in this effect as a result of the different absorbencies of adjacent forms. In addition, as the

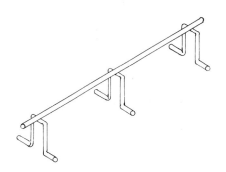

(A) Slab Bolster

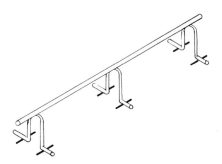

(B) Stainless Dowels on Chairs

(C) Individual High Chair

(D) Stainless Dowels on Chairs

(E) Plastic Protected Legs

FIGURE 3-6 Concrete accessories.

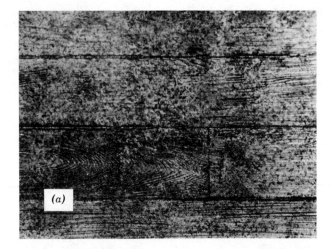

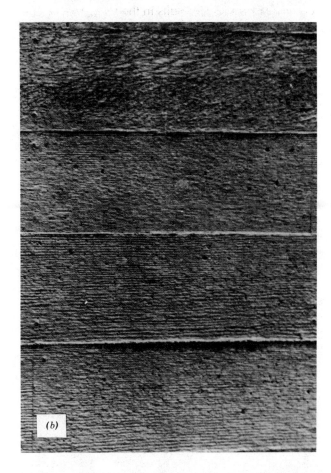

FIGURE 3-7 Off-the-form concrete finishes.
(a) No. 1 Grade T & G surfaced boards.
(b) No. 1 Grade square edged boards. Note leakage at joints.
(c) Slash grain lumber, with surfaced edges drawn together to prevent leakage. (Figure used with permission of Portland Cement Association.)

Lumber

For off-the-form finishes, lumber is used with either a smooth finish such as flooring, or No. 1 grade, T & G surfaced boards, or a textured finish derived from lumber that has been rough sawn, sandblasted, or otherwise distressed to impart a textured finish to the concrete. See Figure 3-7 for lumber used as forms for varying surface results in concrete.

forms are reused, the absorbency decreases and the darkening effect on the concrete surface becomes less. As a result of varying absorption rates and reuse of forms considerable variation in concrete surface color will ensue.

Plywood

For architectural concrete, plywood with plastic overlays (medium or high density) produces a smooth, pleasing concrete surface. Using plywood without this overlay will result in concrete surfaces which reflect graining and boat patches—a telegraphing of the plywood onto the concrete.

A plywood product developed in Finland made of a plastic-coated birch plywood, sometimes called Finn-Fir, is gaining increased popularity. This is due to its availability in larger special sizes, thereby reducing the number of joints, and its better-than-average reuse factor.

Steel

Since steel is impervious, a uniform color usually results from its use. Since steel can be made to special sizes, the number of joints can be reduced. The thickness of the steel form is important to ensure against deflections between support members. To prevent rusting and staining, epoxy coatings, or parting agents with rust inhibitors should be applied. Steel forms can provide between 50 and 100 reuses.

Fiberglass-Reinforced Plastic

Plastic forms usually are a product of cloth fiberglass or random glass fibers and polyester resin formulations. Since the form can be molded to almost any shape and size, jointing can be practically eliminated. Although this form may be the most expensive, the final cost is low because of its almost unlimited reuse, its light weight for erection, and its reduced number of joints.

Form Liners

When off-the-form finishes are desired, form liners applied to the structural forms may be utilized. The form liner will impart its inherent texture to the finished concrete surface. Form liners may be plastic, such as rigid PVC, acrylonitrile-butadiene-styrene (ABS); fiberglass-reinforced polyester; urethane; or rubber mattings.

FORMING

Since placing and consolidation of architectural concrete are more severe, there is a need to ensure that the formwork be more securely braced to avoid bulging, offsets, and similar distortions. This may include the addition of extra back-up members beyond that required for usual structural needs. The requirements of ACI 347 "Recommended Practice for Concrete Formwork" require modification in order to assure more rigid formwork.

Form Joints

Leakage at form joints results in the formation of surface blemishes that cannot be eradicated, even with abrasive blasting. See Figure 3-8 for disfigurement resulting from leakage at joints.

The resulting blemish is characterized by visible aggregate adjacent to normal dense concrete, and a significant color change. The use of liners with joints offset from the joint of the structural forms will reduce leakage. A rustication strip at joints, securely fastened and sealed at the joint, will also minimize leakage. Silicone or other rubber-base sealants used at butt joints will help prevent leakage. Where exposure of the aggregate is desired by blasting, bushhammering, or tooling, a thin, pressure-sensitive tape may be used to prevent leakage at joints. Taped joints should be inspected prior to placing concrete to make certain that the tape has not moved.

FIGURE 3-8 Leakage at joints is not eradicated even with sand blasting.

Form Ties

When designing for architectural concrete, consideration must be given to the type of tie system to be employed, since the selection of a specific type will have an influence on the aesthetics of the finished concrete.

Since each tie system leaves a characteristic hole, the architect must be familiar with the different systems and the profiles of the holes they create.

Ties used in architectural concrete should leave no metal closer than 1½ inches from the surface, and generally fall within the following types:

Snap tie: A stainless steel single member with a positive break-back feature and an optional plastic cone. Without the cone, the snap tie leaves the smallest hole.

Coil tie: An assembly utilizing reusable coil bolts, washers, and cones. Coil ties can be used for reattachment of formwork or false work, and are available in stainless or galvanized steel.

She-bolt tie: A type of tie and spreader bolt where the end fastenings are threaded into the end of the bolt, thereby eliminating cones and the size of the hole. The outer fastening devices are reusable.

He-bolt tie: A type of tie where the outer fastening devices are reusable, with an expendable female threaded unit left in the wall.

All of these ties leave round and relatively clean holes that may be patched flush, recessed, or plugged with cement cones to match surrounding concrete. Where the holes are to be exposed, a representative pattern of the tie layout should be drawn on the architectural drawings and be checked on shop drawings to ensure a consistent arrangement. See Figure 3-9 for concrete form ties.

PLACING AND CONSOLIDATING

The concrete contractor's crew and supervisor are the most important aspect of this phase of the concrete operation. They must be advised as to the special care and attention to be given to the material, which is to be construed as an architectural finish and as such must be handled judiciously from start to finish. A description of the methods and sequence of operations should be submitted by the contractor for approval by the architect and engineer. The preconstruction mock-up and conference should include the concrete contractor's key personnel so that they are impressed with the importance of their involvement.

The scheduling of ready-mix trucks to the site should be coordinated to avoid excessive mixing while waiting or delays in placing. Nonuniform mixing can contribute to nonuniformity of appearance, and delays result in the formation of cold joints.

Trucks used for conveying of architectural concrete should be cleaned after every delivery and should not be used for transporting other mixes to avoid contamination of the specified architectural mix.

Concrete should be placed or deposited (never poured) in uniform layers to prevent segregation and so that vibrators are not used to move concrete into final position. By placing concrete in final position rather than pouring, splatter is avoided on forms that may result in blemishes as it hardens on the form face.

Consolidation of concrete by vibrators is particularly important in architectural concrete to minimize surface voids and to blend lifts of concrete in successive layers to ensure uniformity of appearance. ACI-309, Consolidation of Concrete, provides a detailed recommendation on selection of vibrators and procedures.

SURFACE TREATMENT

Off-the-form finishes are those imparted to the concrete surface on the basis of the form or liner face and are indelibly etched into the surface during the casting and molding stage. Surface treatment implies exposing the fine or coarse aggregate of the hardened concrete after the forms are removed. This can be accomplished by methods such as abrasive blasting (sandblasting), bushhammering, or mechanical tooling. Other means of exposure used less frequently are brushing and washing before the concrete reaches full strength, high-pressure water jet, surface retardation, and acid wash. As noted earlier, surface treatment will affect the appearance of the coarse aggregate and the selection of aggregate to enhance its appearance is paramount.

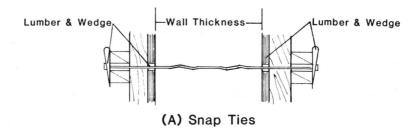

(A) Snap Ties

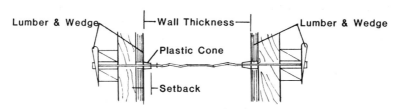

(B) Snap Tie with Plastic Cone

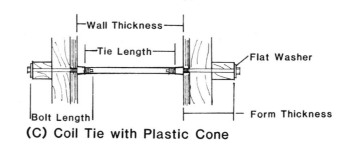

(C) Coil Tie with Plastic Cone

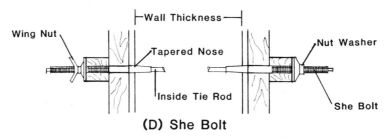

(D) She Bolt

FIGURE 3-9 Representative form ties.

Abrasive Blasting (Sandblasting)

Sand or abrasive materials are used in blasting to reveal the mix ingredients of the concrete to specific depths to achieve certain architectural effects. These effects are illustrated in Figure 3-10 and are classified as follows:

1. *Brush Blast.* A light scouring of the concrete surface is achieved that removes some surface blemishes and slightly exposes the fine aggregate.

2. *Light Blast.* Approximately ¹⁄₁₆ inch depth of reveal, removing the surface skin sufficiently to expose fine aggregate with occasional exposure of coarse aggregate. A uniform color results from the primary influence of the fine aggregate.

3. *Medium Blast.* Approximately ⅛ to ³⁄₁₆ inch depth of reveal, sufficient to expose coarse aggregate.

4. *Heavy Blast.* Approximately ¼ to ½ inch depth of reveal, exposing the coarse aggregate to a maximum projection. A gap-graded aggregate is desirable for heavy blast texturing.

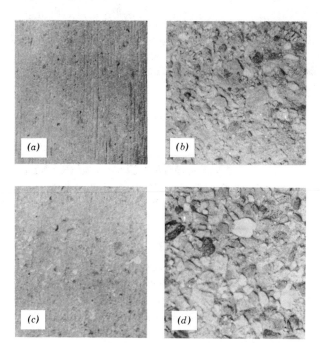

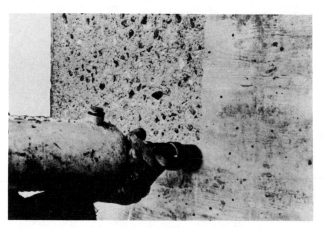

FIGURE 3-11   Pneumatic bushhammer. (Used with permission of Portland Cement Association.)

FIGURE 3-10   Degrees of abrasive blasting finish. (A) brush; (B) medium; (C) light; (D) heavy. (From *Guide to Cast-In-Place Architectural Concrete* Practice (ACI 303R-74) (Rev. 1982) ACI Committee 303 American Concrete Institute.)

Blasting may be performed with abrasive materials such as silica sand (or wet sand, where air pollution standards prevail), aluminum carbide, or black slag particles. Blasting is performed at a time which is dictated by a number of factors, including scheduling, economics, strength of concrete, and hardness of aggregate. Prior trial tests on the mock-up will provide useful information, as will trial tests on basement walls, to establish criteria and techniques.

Generally, the time at which blasting is performed is related to the depth of reveal. The deeper the reveal, the sooner the blasting should be performed after removal of forms.

Bushhammering

Bushhammering is a process in which mechanically or hand-operated hammers remove the skin of hardened cement paste from the surface of the concrete and fracture the coarse aggregate at the face of the concrete to reveal an attractive, varicolored, and textured surface.

Power hammers can be driven electrically or by

compressed air. (See Figure 3-11 for pneumatic bushhammer.) Hand hammers are used for small as well as for restricted locations. By using this technique, the normal thickness of material removed from the face of the concrete is about ⅛ inch; however, by going over the surface more than once, a greater thickness of material can be removed.

Concrete should not be bushhammered until it has attained a strength of at least 3500 psi. Since it is economically unfeasible to do bushhammering until all forms have been removed, it generally cannot be started until three weeks or longer after casting, by which time the desired strength has been achieved.

Since bushhammering reveals the aggregate, the selection of the coarse and fine aggregate is of great importance. As a general rule, crushed-stone coarse aggregate is more suitable for bushhammering than uncrushed gravel. The use of uncrushed gravel can lead to bond failure between the aggregate and the matrix, which may cause some of the aggregate particles to become loose and fall out. The aggregates that behave best under bushhammering are those that can be cut or bruised without fracturing. Most of the igneous rocks, including granites, are well suited for this purpose. So are the hard limestones.

The treatment of arrises also requires careful design. Although one can use hand bushhammers right up to the arris, one should recognize the possibility of inadvertently damaging the sharp corner, thereby requiring repair which will be visible in the completed work. This problem can be dealt with in two ways: (1) by providing chamfered or rounded corners of at least 1 ½ inches radius; (2) by attaching a ¼ inch wood fillet at least 2 inches wide to form a plain margin up to which the tooling can be carried.

Mechanical Tooling

Fracturing of the surface of concrete other than bush-hammering is produced by scaling, chipping, jack-hammering and tooling, using tools appropriate to the texture and configuration desired of the exposed aggregate surface.

Scaling is done with a pneumatic device that rotates and fractures the concrete close to the surface without imparting deep reveals.

Jackhammering is performed with a chiseled or pointed tool at a time when the concrete has attained a high strength so that coarse aggregates are not dislodged during the process.

Tooling produces reeded, striated, and corrugated patterns, depending upon the type of tool used and its orientation as the concrete is worked. Tooling should be kept uniform throughout the work to obtain the desired effect.

When concrete surfaces are fractured by bush-hammering or tooling, the hammers, chisels, and tools must be inspected periodically to ensure that they are not worn to the point where the appearance of the fractured surface is no longer uniform.

PATCHING AND REPAIRING

For off-the-form concrete, patching and repairing obviously follow form removal. For exposed aggregate finishes, patching and repairing are best done after the abrasive blasting or mechanical tooling has exposed the aggregate, since the patch or repair can be damaged through the action of aggregate exposure.

Patching requires matching of the adjacent concrete surfaces in color and texture. Trial mixes of the cement used (gray, buff, or tan) should include some percentage of white cement as well as the original fine and coarse aggregate. The amount of white cement to be used will depend on dried samples that have cured a minimum of 7 days and preferably 28 days. Depending on the size of the patch, the coarse aggregate may either be part of the mix or hand placed after patching. The patching and repair are usually an artisan's task and require diligence and careful quality of work.

Where off-the-form concrete is to be patched, the patch should be compressed into place with the same

form or liner used initially to obtain the same imprint to match adjoining surfaces.

Where necessary, areas to be patched should be cut back to sound material so that the patch will adhere to a sound surface.

ARCHITECTURAL PRECAST CONCRETE

Architectural precast concrete has been utilized since the 1920s, but it was not until the 1950s that its use became more widespread. This has come about through the introduction of more efficient, controlled, and coordinated manufacturing facilites. Such plants usually include organized production, quality control, more reliable quality of work through year-round employment, and the advantages of production in a controlled environment.

Architectural precast concrete has a number of functional advantages. These include incorporation of

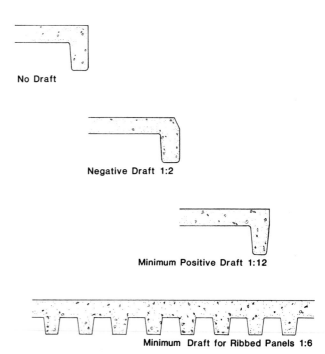

No Draft

Negative Draft 1:2

Minimum Positive Draft 1:12

Minimum Draft for Ribbed Panels 1:6

FIGURE 3-12 Draft concepts for precast concrete. (Courtesy of Prestressed Concrete Institute.)

structure along with aesthetics as in the design of load-bearing elements, the inclusion of sound isolation, thermal insulation, and the ability to erect the elements at the site during all types of weather. Where structural requirements dictate, elements may be prestressed.

Since precast units are a manufactured product, it is essential to work with precasters during the early design development stages in order to get input from the precaster as to the design limitations with respect to manufacturing capabilities, transportation, and erection. The characteristics that precasters can provide information about include draft (the ability to strip the unit from the mold), sizes (limitations on transportation from plant to site), reinforcement (for handling and erection), connections, tolerances, (as-cast and erection), finishes, joint treatment, and the use of master molds to keep costs in line. See Figure 3-12 for typical Precast Concrete Institute guidance on draft requirements.

MATERIALS

The materials used for precast units (i.e., cement, aggregate, coloring pigments, admixtures, etc.) are in general similar to those used in cast-in-place concrete, and with the similar admonition that all materials come from the same source for quality control.

SAMPLES AND MOCK-UPS

It is advisable to visit precasters' facilities and view the various samples, finishes, textures, and aggregates that are available. For major projects, the owner's expenditure of funds for the development of preconstruction samples and mock-ups would be advisable. The mock-up would serve as a jumping-off point to improve the appearance or make modifications to reduce cost without sacrificing the design intent.

Project samples should be at least 18 inches square by 1 ½ inches thick. For major projects, 4 foot square samples should be requested after approval of the 18 inch square samples, especially if exposed aggregate is involved. Mock-ups should be representative of a full-size, typical module.

FINISHES

There is an abundance of finishes for architectural precast concrete, giving the designer a wide choice. Essentially, finishes can be achieved during any one of three distinct phase in the manufacturing process: prior to casting, before hardening, and after hardening.

Prior to Casting

These finishes can be predetermined and obtained using forms and special materials within the forms:

1. *Smooth finishes:* Obtained simply by using nonporous forms such as fiberglass, steel, sealed plywood, overlaid plywood, or sealed concrete. The

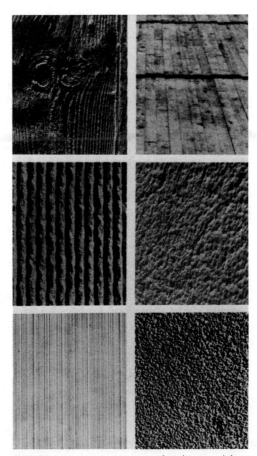

FIGURE 3-13 Various patterned and textured form liners attached to forms to produce precast textured surfaces. (Courtesy of Prestressed Concrete Institute.)

smooth-form surfaces impart a similar surface on the faces of precast architectural concrete.

2. *Textured finishes:* Result from using patterned or textured form liners such as rubber matting, textured fiberglass forms, rough-sawn wood, or any variety of form which has been textured or patterned. See Figure 3-13 for special textured form liners.

3. *Special finishes:* Acquired by placing selected materials other than concreting materials at the bottom of the form. These materials may be ceramic tile, marble, granite, brick, or cobbles. They may be placed so as to obtain a complete facing, or they may be spaced so that a mortar joint is formed between them by the concrete matrix.

Before Hardening

These finishes are induced upon the faces in contact with the forms before hardening during the precasting process:

1. *Chemical retardation:* Retarders are applied to those surfaces of the forms which correspond to the panel faces selected for aggregate exposure. Upon placing the concrete in the forms, the retarders inhibit and slow down the chemical process involved in concrete hardening. The retarded cement paste is then removed by jetting with water and/or brushing. The degree of etching can result in any one of three textured surfaces:

   a. *Light etch.* The outer surface of the cement paste and sand is removed, resulting in only a minute exposure of coarse aggregate, approximately 1/16 inch reveal.

   b. *Medium etch.* A greater amount of the cement paste and sand is removed, resulting in a partial exposure of the coarse aggregate, approximately 1/8 to 3/8 inch reveal.

   c. *Heavy etch.* A significant amount of cement and aggregate is removed, resulting in a uniform appearance of coarse aggregate, approximately 1/4 to 3/4 inch reveal.

2. *Treatment of exposed face:* These finishes are applied to the surfaces of the exposed face of the precast unit while it is still in the plastic state. Such a technique may consist of brooming, stippling, or using a roller with a textured surface to impart the desired texture on the exposed face.

After Hardening

These finishes are obtained after the precast units have been cast and have attained their required strength.

1. *Acid etching.* Light and medium etching, described previously for chemical retardation, may be achieved by brushing the units with acid or dipping in an acid bath.

2. *Abrasive blasting.* Light, medium, and heavy exposure of aggregates may be obtained by blasting the units with sand or an abrasive aggregate. The best possible appearance of the final surface will be achieved by using a gap-graded concrete mix of low slump and adequate cement content. Gap grading, or skip grading as it is sometimes called, omits some of the intermediate sizes of coarse aggregates normally included in standard concrete mix. Gap-graded mixes result in a uniform size distribution of the exposed coarse aggregate. Abrasive blasting may sometimes result in a dulling of the aggregate, including the loss of sharp edges.

3. *Bushhammering.* This is a process in which mechanically or hand-operated hammers remove the skin or hardened cement paste from the surface of the concrete, fracturing the coarse aggregate at the face of the concrete to reveal an attractive varicolored and textured surface. Power hammers are faced with a number of points and are driven by electricity or compressed air. Hand hammers are used for small areas as well as in restricted locations. The precast members should not be bushhammered until they have obtained a strength of about 3750 psi. The aggregates that behave best under bushhammering are those that can be cut or bruised without fracturing. Most of the igneous rocks, including granite, are well suited for this purpose, as are the hard limestones. Since corners can be damaged during brushhammering, it is essential to use either a rounded corner or stay about 1 or 2 inches away from corners.

4. *Honing or polishing.* In this process the faces of exposed units are ground to the desired appearance by mechanical abraders, starting with a coarse grit and ultimately finishing with a fine grit.

Appropriate aggregates to use in exposed aggregate finishes are granite, quartz, quartzite, crystalline aggregates, onyx, pebbles, marble, glass, ceramics, silica sand, and special abrasives such as carborundum and alundum. In using aggregates in concrete it is

FIGURE 3-14 Typical precast panel being hoisted for placement on precast facade. (Courtesy of Prestressed Concrete Institute.)

recommended that a petrographic analysis be made to determine their suitability as a concrete aggregate.

Other textures and finishes are possible. However, it is recommended that procedures can be reviewed with precasting plants to determine feasibility and economics. See Figures 3-14 and 3-15 for illustrative examples of precast concrete buildings.

FIGURE 3-15 Typical precast structure and an insert showing light aggregate exposure using a retarder. (Courtesy of Prestressed Concrete Institute.) This illustration also appears in the color section.

SPECIAL CONCRETING TECHNIQUES

PREPACKED CONCRETE

Prepacked concrete, also known as intrusion or grouted concrete, was developed essentially for placement under water, for mass concrete bridge piers, and for heavy concrete work such as the construction of steam hammer foundations and the restoration of dams. The same principle can be applied to architectural concrete finishes where specially selected aggregate is desired for exposure to obtain aesthetic effects.

With ordinary concrete, cement, fine and coarse aggregates, and water are mixed together before being placed within the formwork. In the prepacked method, the forms are filled with coarse aggregate only; subsequently, the interstices are grouted with a specially prepared mortar. The final stage in the process for architectural concrete is to expose the aggregate by removing the skin of cement by wire brushing, sandblasting, bushhammering, or grinding.

Structural concrete placed in normal construction contains approximately 1950 pounds of coarse aggregate per cubic yard. In prepacked concrete, about 2700 pounds of coarse aggregate per cubic yard can be obtained. This provides approximately 23% more aggregate available for exposure at the surface. In addition, with ordinary concrete, there can be segregation of aggregate so that some aggregate may not be near the surface upon exposure by sandblasting or wire brushing.

The materials for prepacked concrete should all be obtained from the same source in order to be of consistent quality and color; otherwise, variations in material will be reflected in the finished product.

The formwork for prepacked concrete must be of a high standard to prevent deformation and the loss of grout at joints, due to the internal pressure of the expanding mortar. To prevent loss of grout, joints in the formwork should be sealed.

The operation of placing the course aggregate can be assisted by the use of form vibrators. Generally, the coarse aggregate is placed in horizontal layers in a minimum void content. The selection of coarse aggregate is the choice of the architect; it may range in size from ⅜ inch to 3 inches. The aggregate should be washed thoroughly and well drained of water before use; care should be taken to prevent it from becoming contaminated while on the site. Generally, the course aggregate should not be dropped into the forms from heights greater than 5 feet.

The mortar used for prepacked concrete is usually an activated cement mortar having an intrusion aid. The intrusion aid contains a dispersing agent and a chemical that reacts with the alkalis of the cement to form a gas. Its action improves the workability of the grout, reduces bleeding, and causes a slight expansion before final setting.

The mortar is placed during the grouting process through 1 inch diameter steel pipes, placed vertically and spaced about 2 feet O.C. or as necessary to insure uniform placement of grout. Initially, the outlet end of the pipe is about 2 inches above the bottom of the formwork. The pipes are raised as the grout rises. After grouting has progressed sufficiently, the pipes should extend to a depth of 12 inches in the grout. In order to check the level of the grout in the forms, holes are sometimes bored in the forms that can be plugged later. Another means of checking grout levels is to install small lights of acrylic plastic in the forms so that the grout may be seen. Forms are generally vibrated during the grout intrusion process to help distribute the grout uniformly.

The sandblasting of the surface to expose the aggregate should be carried out as soon as possible after the forms have been stripped.

CYCLOPEAN CONCRETE

Cyclopean concrete is an example in which prepacked concrete is used to obtain a unique architectural effect with boulder-size aggregate.

Concrete walls using exposed granite aggregate from 4 to 8 inches in size have been erected for dormitories at Yale University. The following sequence was used in the construction of 12-inch thick bearing walls:

The back form was erected to a full 8 feet height, and the front form was erected in 2-foot increments to facilitate placing the stone. The granite was tipped into the forms from wheelbarrows, but was hand-placed where necessary.

The forms were coated with a retarder to delay the set of the grout surface. Two-inch diameter grout pipes, placed 4 feet on centers, extended to the bottom of the forms.

A 1:2 grout of soupy consistency was pumped into the forms to fill the voids in the stone, and the grout pipes were slowly raised as the grout rose in the forms. The lower ends of the pipes were kept below the surface of the grout, and the grout elevation was determined by tapping the forms. The forms were stripped the following day, and the grout was scraped off the surface of the stone. See Figure 3-16 for typical cyclopean concrete test panel.

FIGURE 3-16 Test panel of cyclopean granite placed by prepacked method.

SLIP FORMING

Slip forming is a technique using sliding forms that move continuously and shape the concrete as they slide over the freshly placed mix. Vertical slip forms have been used to shape silos, storage bins, and chimneys for industrial uses, but have been adapted architecturally to shape elevator towers, stair towers, and vertical enclosing elements of buildings.

Steel forms are used and move at a rate of about 6 to 12 inches per hour, with the process continuing around the clock, and with the forms moving so that they leave the formed concrete only after concrete is strong enough to retain its shape and support its own weight. The procedure is akin to an extrusion process.

A jacking system raises the forms and the working deck assembly, which in turn are supported on steel rods embedded in the concrete. Since it is a 24-hour operation, the logistics for supplying the concrete and manpower require careful planning.

Although a smooth finish is cast, the hardened surface can be treated by abrasive blasting or bushhammering to obtain the required exposed aggregate texture. Design mixes incorporating special aggregates to be exposed may be used as in any architectural concrete work.

TILT-UP CONSTRUCTION

In the tilt-up system, panels of concrete are cast in a horizontal position, on a flat, smooth and level bed with a parting agent between the bed and the panel. Both load-bearing and non-load-bearing wall panels may be cast in this manner with required reinforcement, and if needed with prestressing strands.

The panels are then rotated about their bottom edges into a vertical position and secured to the structure through connections embedded in the panels.

For architectural purposes, exposed aggregate surfaces can be obtained by using the sand embedment method. In this method, a layer of fine sand is spread over the bottom of the form to a depth of about one-third the diameter of the aggregate. Pieces of aggregate are pushed into the sand to obtain the densest possible coverage.

FIGURE 3-17 Precast panel being tilted up into position. (Used with permission of Portland Cement Association.)

After the aggregate is in place, water is sprayed to settle the sand around the aggregate so that each piece is embedded securely to a depth of about one-half to one-third the diameter.

Steel reinforcement is placed on plastic or galvanized chairs, and the concrete is carefully placed over the embedded aggregate.

As an alternate method, the aggregate can be placed on the form and then sprinkled with fine, white, dry sand to a depth of one-half to one-third the aggregate.

After proper curing, the panels are raised, and any sand clinging to the exposed aggregate is removed by brushing, airblasting, or washing with water. See Figure 3-17 for typical tilt-up construction.

ARCHITECTURAL FORMS

Structural concrete using normal or lightweight aggregates with off-the-form finishes is used to obtain unusual shapes due to the plasticity of concrete. These shapes are achieved through cast-in-place procedures or precast and prestressed techniques to develop unusual architectural forms.

CONCRETE SLABS

The simpler shapes are those used for floor-ceiling concrete slabs as follows:

Flat Slab

A concrete slab reinforced in two or more directions, generally without beams or girders. The supports are columns with dropped panels which pick up the floor loads. Generally this construction is used for bays that are almost square and require heavy loading, as in warehouses, garages, and so forth. See Figure 3-18 for flat slab construction.

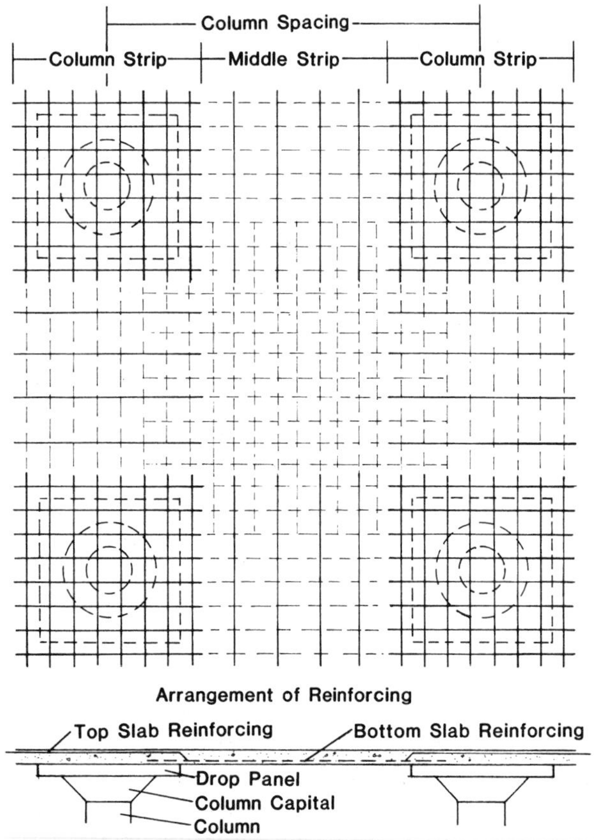

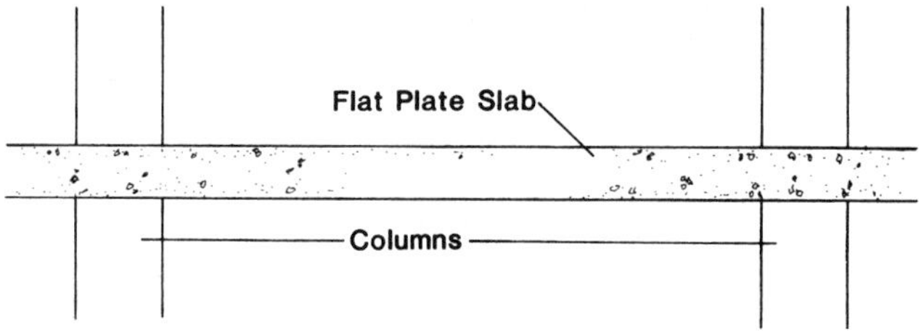

FIGURE 3-19 Flat-plate floor construction.

Flat Plate

Where lighter floor loads are encountered, as in apartment buildings, office structures, dormitories, and so forth, a modified flat slab is used without drop panels. This type of construction provides a flat continuous ceiling without breaks, making for easier subdivision of like spaces. See Figure 3-19 for flat plate construction.

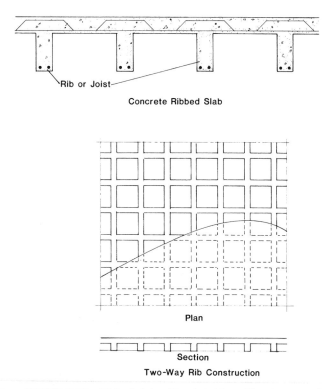

FIGURE 3-18 Two-way flat slab. (Reprinted from Huntington and Mickadeit, *Building Construction,* 5th Ed., New York: John Wiley & Sons, Inc., 1981.)

FIGURE 3-20 Ribbed slabs. (Reprinted from Huntington and Mickadeit, *Building Construction,* 5th Ed., New York: John Wiley and Sons, Inc., 1981.)

Ribbed Slab

A ribbed slab is a reinforced concrete panel consisting of a thin slab reinforced with a series of ribs. When the ribs are in two directions it produces a waffle pattern. See Figure 3-20 for ribbed slab construction.

SHELLS, ARCHES, AND DOMES

Unusual shapes are obtained through casting or precasting and require competent structural engineers to work with the architectural designer to achieve a wide multitude of thin shell constructions. Thin shell construction is characterized by relatively thin slabs and web sections as follows:

Folded Plate

This construction consists of thin, flat elements of concrete, connected together to form a series of triangles when viewed from the end (similar to accordion folds), and capable of carrying a load over a long span. See Figure 3-21 for folded plate construction.

Arches

The curved arch or barrel arch forms an opening width (span) and a height (rise) resembling a barrel with a

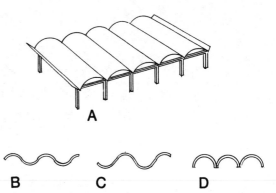

FIGURE 3-22 Barrel shell roof. (Reprinted from Huntington and Mickadeit, *Building Construction*, 5th Ed., New York: John Wiley & Sons, Inc., 1981.)

constant cross section. Barrel arches are used extensively for long spans as roofs over armories, gymnasia, field houses, and so forth. See Figure 3-22 for barrel shell roof.

Domes

Where circular structures are designed, such as gymnasia, field houses, and auditoria, the accompanying roof is usually a dome shaped structure. The dome acts as a membrane and has a tension ring at its major circumference and a compression ring at the top of the dome. Some domes are often constructed using a mound of earth as the form, then lifted into final position after the concrete has set. See Figure 3-23 for circular dome roof.

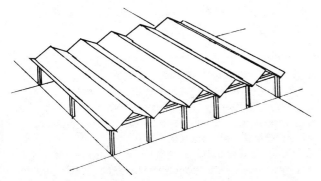

FIGURE 3-21 Folded plate roof. (Reprinted from Huntington and Mickadeit, *Building Construction*, 5th Ed., New York: John Wiley & Sons, Inc., 1981.)

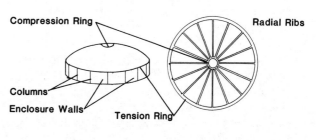

(A) Dome   (B) Pattern with Radial Ribs

FIGURE 3-23 Circular dome roof. (Reprinted from Huntington and Mickadeit, *Building Construction*, 5th Ed., New York: John Wiley and Sons, Inc., 1981.)

PROTECTION OF ARCHITECTURAL CONCRETE

Architectural concrete, either precast or cast-in-place, often requires a clear protective sealer to overcome a number of concerns as follows: (1) temporary mottling and darkening of the surface after a rainstorm; (2) accretion of dirt and soot on exposed aggregate surfaces; (3) change in color of the matrix due to attack by atmospheric pollutants; and (4) the etching of gray and bronze glass below when alkalis leach from the concrete onto the glass.

To overcome these problems, one may use an application of clear coating to preserve the initial appearance of exposed concrete. Generally, the preferred coating should be clear, water repellent, and a breathing type. These coatings function primarily as follows: (1) by reducing water penetration, they minimize mottling and darkening due to rainstorms; (2) by reducing water penetration, free alkali in concrete is not leached out onto the concrete; (3) they tend to make the surface self-cleaning, reducing the accumulation of dirt; (4) they tend to reduce atmospheric attack on the cement matrix; and (5) the breathing qualities of the coatings provide for uniform weathering.

Many coating types have been marketed by manufacturers as a panacea for the problems outlined above. However, they are not without their own contributory problems. Some have a relatively short life span. Some actually attract soot. Some develop a glossy appearance that changes the architectural effect. Some darken the exposed aggregate and the cement matrix considerably, thereby altering the architectural appearance.

The ideal coating should be clear, nondiscoloring, long lasting, and should not make any discernible change in the color of the aggregate or in that of the matrix.

The Portland Cement Association investigated a wide variety of clear coatings, and published its findings in Bulletin D137, 1968, "Clear Coatings for Exposed Architectural Concrete." Sixty products were investigated to determine their effectiveness in protecting exposed aggregate concrete and smooth concrete surfaces against the elements. These coatings consisted of acrylics, polyurethanes, polyesters, silicones, waxes, epoxies, styrenes, and in some products, mixtures of these chemical formulations.

In the accelerated tests, the PCA found that by and large, the coatings based on a methyl methacrylate formulation provided better protection on exposed surfaces than other types of coatings. The laboratory test results were confirmed by the outdoor weathering exposure. Similarly, on smooth concrete surfaces, methyl methacrylate coatings with a higher solids content gave better protection. Generally, the polyurethanes, polyesters, and epoxies tended to cause a glossy appearance and created a yellowing or darkening effect.

PERFORMANCE REQUIREMENTS

When selecting, designing, specifying and constructing in concrete, it would be advantageous to refer to Chapter 1 to ensure that the pertinent performance requirements are reviewed. An example for concrete would be the following analysis.

STRUCTURAL SERVICEABILITY

Concrete is excellent in compression and can be designed for tension loads by conventional reinforcement or by prestressing. Prestressing may also be used to introduce stresses in the material by opposing stresses from the service loads whereby the tensile stresses may be practically eliminated.

For architectural concrete, rather than being simply an appendage, the concrete should be designed for load bearing or other structural functions to reduce the cost of construction, by serving both as the structural element and the architectural element.

FIRE SAFETY

Concrete is not combustible; therefore it does not develop any of the products of combustion associated with fire. However, concrete should be designed for the required hours of fire rating based on ASTM E119, and also by rational design taking into account aspects of fire resistance such as heat transmission, flame penetration, and the maintenance of structural capability. Where penetrations or joints occur in the concrete, there are now materials available such as incombustible foam and gasket materials to provide adequate hourly fire resistance ratings for the penetrations and joints.

HABITABILITY

Thermal Properties

While concrete does not possess insulating qualities, its bulk does provide it with a heat or cold reservoir since it gives off heat (or cold) rather slowly. Consideration should be given to the use of insulation on the inside face of walls, or below floors of exposed overhangs. In precast wall panels insulation can be introduced between the inner and outer wythe.

Acoustic Properties

Noise transmission through concrete walls is reduced due to its bulk and density. For floors however, impact or tapping noises require sound deadening through the judicious selection of floor covering materials. Stair elements may have neoprene bearing pads at supports to reduce sounds from impact loads.

Water Permeability

Properly designed and constructed, concrete is usually impervious to water. However the introduction of waterproofing, dampproofing, waterstops or other design elements should be considered by an investigation of the proposed design and use.

Hygiene, Comfort, Safety

Walks, steps and ramps subjected to water from rain should have broomed finishes or applications of alundum or carborundum to prevent slipping. Exposed concrete walkways or stairs on interiors should be hardened through proper curing to inhibit dusting.

DURABILITY

While structural concrete in foundations may be subjected to attack by sulfates in the soil and concrete used for work in sea water may be subject to attack by saline salts our concern is for the durability of exposed architectural concrete. The durability of such concrete exposed to the weather may be subject to freeze–thaw cycling and may be affected by chemical de-icing salts.

Environments may be harsh in nature and may exert internal forces on concrete containing moisture. The use of deicing chemicals heighten the problems by increasing these forces.

ACI Standard 301, Specifications for Structural Concrete for Building, requires the use of air entrainment and a limitation on the water–cement ratio to control durability.

COMPATIBILITY

Differing materials may interact with one another as outlined in Chapter 1. In the first instance the ingredients used to make concrete must be compatible. Since new coarse aggregates often are used for the first time to obtain a unique architectural effect it is essential to run a petrographic test to insure that the aggregate is suitable for use in concrete.

In addition, the detailing of concrete on a drawing may be such that rain water drains from a concrete surface onto a bronze or gray solar tinted glass. The alkalies leached from the concrete may etch the glass vividly as a result of the detailing.

One must be alert to all of those possibilities, and a check for compatibility with adjacent materials must be made to assure a problem-free solution.

4

MASONRY

"Masonry" has been generally defined as an assembly of brick, stone, concrete masonry units, structural clay tile, architectural terra cotta, glass block, gypsum block, or similar material bonded together with mortar to form walls and other parts of buildings. However, since modern-day technology is outpacing the redefinition of terms, it might be more practical, for our purpose, to discuss and review these materials by dividing them into two basic categories: (1) unit masonry, and (2) stone.

Unit masonry materials are primarily manufactured units of a size that are generally handled and erected by one mason. However, there are exceptions to this simple definition in that prefabricated panels of brick and concrete masonry units require several masons to erect them. Unit masonry includes such materials as:

Brick	Architectural terra cotta
Concrete block and concrete brick	Ceramic veneer
Structural clay tile	Glass block
Structural facing tile	Gypsum block
	Adobe brick

Stone is primarily a natural quarried material that can be assembled with mortar or which in modern-day technology can be erected and installed with metal anchors and fasteners and jointed with sealants, without the use of mortar. Stone materials that are used for building applications include:

Granite
Limestone
Marble
Sandstone
Slate
Travertine

TABLE 4-1
BRICK SIZES

	Width (in.)	Height (in.)	Length (in.)
Standard	3¾	2¼	8
Modular	3¾	2½	7⅝
Roman	3¾	1⅝	12
Norman	3¾	2¼	12
Econo	3⅝	3⅝	7⅝
Jumbo	3⅝	7⅝	7⅝

BRICK

SIZES

Brick is manufactured in a number of sizes. Few manufacturers make all sizes, and new sizes are added and less popular ones are removed as design requirements change. For example, with the advent of high-rise apartments, econo and jumbo sizes were introduced to provide a face size of a scale consistent with the scale of the building, to reduce the cost of masonry construction, and to keep masonry competitive with other facing materials. Some of the more common sizes are shown in Table 4-1.

Since size and availability vary in many localities, it would be prudent to ascertain this information before proceeding with a design predicated on these aspects.

MANUFACTURE

Brick is made from clay or shale that is finely ground, mixed with water, molded to the desired shape, dried, and fired in a kiln. Three processes are used in the manufacture of brick today.

Stiff-Mud Process: About 12 to 15% water is added to the clay. After thorough mixing, the mixture is extruded through a die, and cut with a tightly stretched wire.

Soft-Mud Process: About 20 to 30% water is added to the clay to produce a uniform plastic mass. The mixture is placed in molds by hand or machine. To prevent the mix from sticking to the molds, the molds are either wetted or sanded. If water is used as the lubricant the brick is called "water-struck"; if sand is used it is called "sandstruck."

Dry-Press Process: Using clays of low plasticity and the addition of up to about 10% water, the mixture is placed in steel molds and subjected to pressures of 500 to 1500 psi, producing the most accurately formed brick.

SPECIAL SHAPES

Where special shapes are required to fulfill a design requirement, as for moldings, completing an unusual opening, or for filling circular columns, consult brick manufacturers being considered for the project to determine whether they can produce the shapes required or the modifications to be made to the design to stay within the limitations of their manufacturing capabilities.

Quite often brick can be cut or clipped to fit certain conditions of the design. However, more refined shapes may have to be specially molded. See Figure 4-1 for special shapes. Color variation may occur due to separate firing of special shapes. Consult with manufacturers and obtain samples to verify color variation.

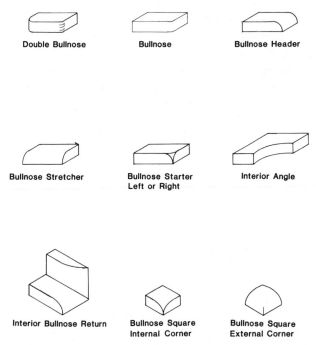

Double Bullnose   Bullnose   Bullnose Header

Bullnose Stretcher   Bullnose Starter   Interior Angle
                       Left or Right

Interior Bullnose Return   Bullnose Square   Bullnose Square
                          Internal Corner   External Corner

FIGURE 4-1 Special brick shapes. (Reprinted from Hornbostel, *Construction Materials,* New York: John Wiley & Sons, Inc., 1978.)

BUILDING BRICK

Formerly called "common brick," building brick is manufactured to meet the requirements of ASTM C62. In many instances building brick has been utilized intentionally as facing brick, depending upon its color after firing and whether its face has been textured by scoring, combing, wire cutting or otherwise treated to provide a unique surface appearance.

Building Brick is classified under ASTM C62 as to durability in accordance with the following criteria:

Grade SW: Severe weathering for exposure to heavy rainfall and freezing.
Grade MW: Moderate weathering for exposure to average moisture and minor freezing.
Grade NW: Negligible weathering for exposure to moisture and freezing.

A Weathering Index Map of the United States shown in Figure 4-2 indicates those areas subject to severe, moderate, and negligible weathering.

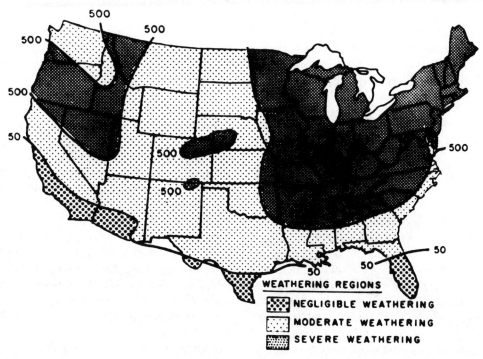

WEATHERING REGIONS

NEGLIGIBLE WEATHERING
MODERATE WEATHERING
SEVERE WEATHERING

FIGURE 4-2 Weathering Index Map of United States, shows where brick masonry is subject to various types of weathering. Reprinted with permission from ASTM, 1916 Race Street, Philadelphia, PA

TABLE 4-2
BUILDING BRICK APPEARANCE TYPES

Type	Use
FBS	For general use in exposed exterior and interior masonry walls and partitions where wider color ranges and greater variation in sizes are permitted than are specified for Type FBX.
FBX	For general use in exposed exterior and interior masonry walls and partitions where a high degree of mechanical perfection, narrow color range, and minimum permissible variation in size are required.
FBA	Brick manufactured and selected to produce characteristic architectural effects resulting from nonuniformity of the individual units.

Building brick, the commonplace red brick, is widely used for foundations and for back-up for exterior solid walls. It is also used as a facing brick when its face texture and color are desirable.

FACING BRICK

Facing brick is made from clay, shale, fire clay, or mixtures thereof, under controlled conditions. Its quality is governed by ASTM C216, which sets forth standards as to durability (weathering) and appearance. Durability consists of two grades, SW and MW, as explained under Building Brick. Appearance is governed by three types, described in Table 4-2.

GLAZED FACING BRICK (GLAZED BRICK)

Glazed facing brick is made from a combination of clay, shale, fire clay, or mixtures thereof similar to the requirements set forth for facing brick, and with a finish consisting of a ceramic glaze fused to the body at above 1500°F. Many colors, textures, and degrees of sheen are available from various manufacturers. When glazed facing brick is intended for exterior use, consult the manufacturer as to suitable material.

Units are produced to meet requirements of ASTM C126. One of the requirements governs grades for tolerances of face dimensions as follows:

Grade S (select): For use with comparatively narrow mortar joints.
Grade SS (select size or ground edge): For use where variation of face dimension must be very small.

The other requirements are concerned with the properties of the finish with respect to imperviousness, resistance to fading, resistance to crazing, flame spread, toxic fumes, and hardness and abrasion resistance.

BRICKWORK

Brickwork involves the assembly of masonry units including mortar, jointing, bond, tieing and workmanship to ensure performance, appearance, strength, and weathertightness. Other factors which influence heat transmission, sound transmission, and fire resistance are a function of the design of a masonry wall.

Mortar

The primary function of mortar is to bond masonry units together so that the mortar joint is durable and acts as a seal against the entrance of water. Other important properties of mortar are workability, water retentivity, and strength.

The primary ingredients of mortar are cementitious material, aggregate, and water. Two standards for mortar are available: ASTM C270 and Brick Institute of America (BIA) M1-72. Essentially, the BIA standard restricts the cementitious ingredients to portland cement and lime, whereas the ASTM standard also allows the use of masonry cement as one of the cementitious ingredients. Canadian Building Digest 163 states: "It is difficult to predict the properties of masonry cements since their composition is not always published. Their use should therefore be based on the basis of known local performance."

BIA standard M1-72 lists four types of mortar as shown in Table 4-3.

Portland cement-lime mortars are proportioned by volume in accordance with information shown in Table 4-4.

TABLE 4-3
MORTAR TYPES

Type	Compressive Strength	Use
M	2500 psi	A high-strength mortar suitable for general use and recommended specifically where maximum masonry compressive strength is required or for masonry below grade and in contact with earth.
S	1800 psi	A mortar suitable for general use and recommended specifically where high lateral strength of masonry is desired.
N	750 psi	A medium-strength mortar suitable for general use in exposed masonry above grade and recommended specifically where high compressive and/or lateral masonry strength are not required.
O	350 psi	A low-strength mortar suitable for use in non-load-bearing walls of solid untis, interior non-loading partitions of hollow units, load-bearing walls of solid units in which the axial compressive stresses developed do not exceed 100 psi and where the masonry wall will not be subject to severe weathering.

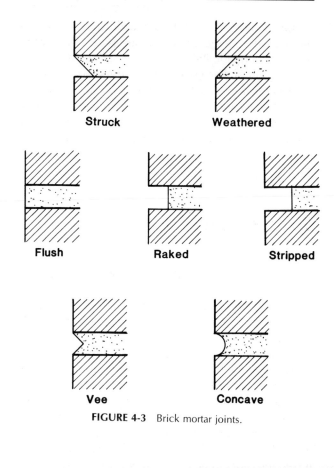

FIGURE 4-3   Brick mortar joints.

Joints

As stated at the outset, the most important functions of mortar are its bonding or adhesive qualities and its durability. Bonding is increased by:

1. Mixing mortar to the maximum flow (ASTM C109) compatible with workmanship. This means using maximum amount of water and retempering.

2. Wetting clay units whose suction rate exceeds 20 grams per minute.

The aesthetics of a brick joint are determined by its size, color, and the manner in which it is tooled. The method by which a joint is tooled also contributes to the watertightness of the wall. The tooled concave joint which compresses the mortar tightly against the masonry units produces the best resistance to water penetration by densifying the surface of the mortar. Joints should be made after the mortar has received its initial set; this compensates for any initial shrinkage.

TABLE 4-4
MORTAR PROPORTIONS BY VOLUME

Type	Portland Cement	Hydrated Lime	Sand Measured Damp Loose Condition
M	1	¼	Not less than 2¼ and not more than 3 times the sum of the volumes of cement and lime used.
S	1	½	
N	1	1	
O	1	2	

Joints that are made by cutting with a trowel or that are rubbed or designed so that they do not shed water run the risk of admitting water. Such joints where desired as part of the aesthetic purpose should be limited to designs using cavity wall construction.

Joint sizes are a function of the masonry unit size and type. Glazed brick joints are generally ¼ inch; facing brick joints, ⅜ to ½ inch; and building brick joints ½ inch. Typical joints are shown in Figure 4-3.

Bond Patterns

While masonry bonding is the laying of units in rows or courses to tie the units together, bonds are also designed to enhance the appearance of a masonry wall. Depending upon the wall design (i.e., solid wall, cavity wall, or faced wall), either masonry bonding or metal ties are employed to bond the units into a solid mass. Each course of brick is one continuous horizontal layer bonded with mortar. The course can consist of brick laid end to end (stretcher) and/or with headers (short dimension). In a solid or cavity wall, each continuous vertical section of masonry one unit in thickness constitutes a wythe. The manner in which the wythes are bonded or tied creates a bond pattern. See Figure 4-4 for bond patterns.

Stretcher or Running Bond: A pattern created by laying brick end to end (long dimension) with each course breaking joints at the midpoint of the course below.

Common Bond: A pattern consisting of stretcher or running bond courses, six or seven courses high with a course of headers (short dimension) laid perpendicular to the stretcher course and thus bonding into the inner wythe.

English Bond: A pattern consisting of alternating courses of stretchers and headers.

Flemish Bond: A pattern created by using one header followed by one stretcher in a course and with each course offset so that a header in one course is centered over the stretcher below.

Stack Bond: A pattern of brick stretchers laid so that horizontal and vertical joints are all in line.

Soldier Course: Brick laid on end with the face showing; used essentially for belt courses or flat arches.

Rowlock Course: Brick laid on face edge with end showing; used for sills or belt courses.

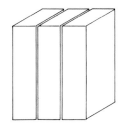
SOLDIER COURSE

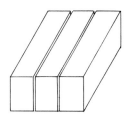
ROWLOCK COURSE

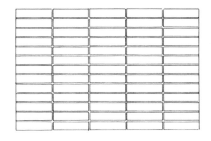
STACK BOND

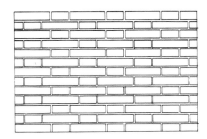
FLEMISH BOND

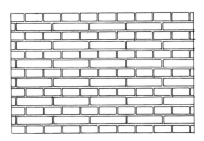
ENGLISH BOND

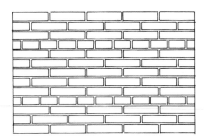
COMMON BOND

FIGURE 4-4 Brick bond patterns and courses.

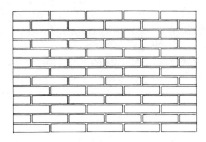

RUNNING OR STRETCHER BOND

FIGURE 4-4   (continued)

Ties and Anchorage

Masonry walls of the solid type are bonded together with either brick bonds or metal ties. Cavity walls and faced walls (against concrete or brick veneer) are tied to the backing with various types of metal ties. Intersecting walls may be bonded with masonry or metal ties.

Metal anchors and ties are usually of zinc-coated steel or other noncorrodible metal, and include a va-

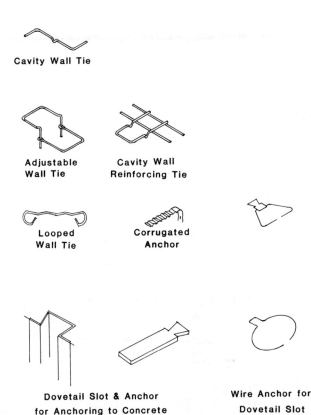

Cavity Wall Tie

Adjustable Wall Tie   **Cavity Wall Reinforcing Tie**

Looped Wall Tie   **Corrugated Anchor**

Dovetail Slot & Anchor for Anchoring to Concrete   **Wire Anchor for Dovetail Slot**

FIGURE 4-5   Masonry wall ties.

riety of types such as wire mesh, wire, corrugated metal, dovetail slots with flat dovetail anchors, rigid steel straps, cavity-wall anchors, and continuous horizontal reinforcement. See Figure 4-5 for masonry wall ties.

A study made by the Armour Research Foundation in 1960 concluded that continuous horizontal reinforcement is as effective as brick headers in tying walls against lateral wind loading. A similar study noted that solid walls bonded with metal ties had significantly greater resistance to water penetration than masonry-bonded walls.

Quality of Work

Leaks in masonry walls are the result of the penetration of water through openings between mortar and brick rather than through mortar or brick. Therefore workmanship, rate of absorption of masonry units, and water retentivity of mortar are the controlling factors affecting the construction of watertight masonry walls. To achieve this end result, do the following:

1. Lay brick in a full bed of mortar without furrowing.
2. Head joints of stretcher courses should have end of each unit fully buttered with mortar.
3. Header courses should have each side fully buttered with mortar.
4. Shove each brick into place so that the mortar oozes out at the top of joints.

In solid masonry construction, lay heavy back-up units first, parge its outer face, then lay face units. Tool face joints as hereinbefore noted under Joints.

Since units with high rates of absorption will suck the water from the mortar, reduce the bond, and induce shrinkage cracks it may be necessary to wet the units prior to laying. Brick with high suction rates can be determined as follows: Draw a 1-inch diameter circle with a wax crayon on the bed of the unit. Place 20 drops of water inside the circle. If the water is absorbed in less than 1½ minutes, the units should be wetted prior to laying, since their suction rate is high.

Water retentivity of mortar is a measure of the flow and workability of a mortar. It is also the property of a mortar which prevents "bleeding" or "water gain" when the mortar is in contact with relatively impervious units. To ensure the proper water retentivity, specify that the mortar have a flow after suction of not less than 75% of that immediately after mixing as determined by ASTM C91.

Efflorescence

The white soluble salt that sometimes appears as a deposit on masonry is known as efflorescence. This deposit is due to the entrance of water into the brickwork, the dissolving of salts (primarily sodium and potassium carbonates and sulfates), and their migration to the outer surface and deposition there in the form of a white soft powder.

All of the ingredients used in masonry construction (brick, mortar, and water) may contain the salts described above. Face brick, ASTM C216 refers to test method ASTM C67 to preclude using brick that may cause efflorescence. Therefore, mortar ingredients and water may be the unknown source of these salts. The use of lime and portland cement of low alkali content will greatly reduce the capacity of mortar to contribute to efflorescence. Water and sand can be checked by laboratory analysis for alkalinity to reduce the possibility of efflorescence.

MASONRY WALL DESIGN

Individual masonry units have certain physical properties relating to compressive strength, flexural strength, and water absorption. Mortar also has certain physical characteristics such as bond strength and tensile strength. While the properties of masonry walls are affected by the physical characteristics of the individual units and mortar, they are also influenced to some degree by other factors, such as quality of work and bonding, which have a bearing on lateral loading, water resistance, thermal properties, etc.

Solid Masonry Walls

Masonry walls may be constructed using two or more wythes of solid brick or by using an outer wythe of brick and concrete masonry unit back-up. Prior to the current increased use of concrete masonry units as a back-up unit, hollow clay tile was used. The masonry wythes may be tied together using brick bonding or metal ties. The spacing of brick bonding units or metal ties is governed generally by local building codes.

Glazed face brick should not be used in solid masonry wall construction. Since water may enter a completed masonry wall, it must have an opportunity to escape, generally by evaporation from the face. When glazed brick is used in solid wall construction, the face glaze may pop off from the forces exerted by the pressure build-up of the water in the form of water vapor.

Cavity Wall Construction

The cavity wall has been used in the construction of many types of structures in Great Britain and Europe for at least 100 years, primarily because of its resistance to rain and water penetration. This resistance to water penetration is based on its unique air space that serves as a break in the transmission of water from the outer face of the exterior wythe to the inner face of the interior wythe.

A cavity wall consists of two walls or wythes separated by an air space (generally 2 inches) and joined together by means of metal ties. Three elements are essential to ensure the success of a cavity wall: (1) drainage weep holes at its base to release water entering the system; (2) corrosion-resistant ties to unite the wythes structurally; and (3) a cavity free of mortar or other material that might form a bridge by which water can migrate to the inner wythe.

Additional advantages over solid wall construction include reduced weight, increased thermal efficiency, and protection against water infiltration into the structure. While water may enter the exterior wythe, it will run down the inside face of the exterior wythe at a flashing gutter and then exit out at weep holes, as shown in Figure 4-6.

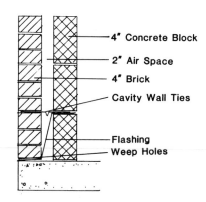

FIGURE 4-6 Typical masonry cavity wall.

Veneered Walls

A veneered masonry wall is one in which an outer wythe of masonry is tied to a structural backing. It is most commonly used in conjunction with wood-frame construction, with an air space between the masonry and the sheathing and tied to it with noncorrosive metal ties. Veneered masonry may also be used as an architectural finish where structural concrete is used. It has also been adapted for use with an exterior framing system of metal studs and gypsum sheathing, veneered with brick secured to the system with metal ties. See Figure 4-7 for a typical veneered masonry wall.

FIRE RESISTANCE

One of the more important provisions of building codes is the regulation governing fire safety. How walls, floors, columns, and other building elements perform in the presence of fire conditions is of major concern to the architect and the public.

The fire resistance of walls, partitions, columns, and other enclosures is based upon the ASTM standard for Fire Tests of Building Construction and Materials E119. While it does not measure the fire hazard in terms of actual performance in a real fire situation, it gives a comparison of the measure of performance between assemblies tested under similar conditions.

Various configurations of masonry construction will develop certain fire-resistance ratings. Tables 4-5 and 4-6 illustrate fire resistance ratings for typical load-

TABLE 4-5
LOAD-BEARING BRICK WALLS

Wall Thickness (in.)	Wall Type	Ultimate Fire Resistance (hr.)
4	Solid	1¼
8	Solid	5
12	Solid	10
10	Cavity	5

TABLE 4-6
FIREPROOFED COLUMN

Construction	Rating
Steel Column, 6 x 6 inches or larger	
3¾-inch brick with brick fill	4 hours
2¼-inch brick with brick fill	1 hour

TABLE 4-7
U-VALUES AND R-VALUES—EXTERIOR MASONRY

Wall Type	U-Value	R-Value
Solid		
8-inch brick	0.48	2.1
12-inch brick	0.35	2.9
Cavity		
10-inches (2-inch air space)	0.33	3.0

bearing walls and for a fireproofed column, respectively.

When designing for fire resistance requirements, check pertinent local building codes, and the Fire Resistance Ratings published by the American Insurance Association.

Thermal Characteristics

The characteristics concerning heat flow are explained in Chapter 1, under the heading Thermal Properties—Thermal Transmittance and Resistance.

The thermal transmittance of masonry walls may be expressed in U-values and the thermal resistance in R-values, as shown in Table 4-7. The R-values or thermal resistances shown in Table 4-7 are quite low or negligible as compared to current requirements limiting heat loss through exterior walls. However, by adding insulation to the back of exterior solid walls,

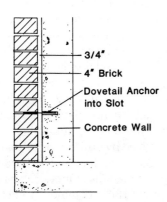

3/4"

4" Brick

Dovetail Anchor into Slot

Concrete Wall

FIGURE 4-7 Typical veneered masonry walls.

TABLE 4-8
STC VALUES FOR VARIOUS MASONRY WALLS

STC Values	Construction
45	4-inch brick, solid
50	10-inch brick, cavity wall, 2-inch air space
52	8-inch brick, solid
59	12-inch brick, solid

or by installing insulation in the air space of cavity walls, increases in R-values can become quite significant, depending on the type and thickness of insulation.

Sound Insulation

The sound transmission loss of a wall or partition is that property which enables it to resist the passage of sound from one side of the construction to the other. See Chapter 1 and the discussion on Acoustics for more definitive explanations.

Insofar as sound absorption is concerned, brick masonry is a poor absorber of sound energy. However, there are perforated and/or slotted structural glazed facing tiles that do have the capacity to absorb sound energy; this material is discussed later in this chapter.

The sound transmission loss of a number of brick masonry walls together with their sound transmission class (STC) and construction are shown in Table 4-8.

Expansion and Contraction

Masonry walls contract and expand with changes in temperature and this factor must be taken into account during the design of the structure. The average coefficient of thermal expansion for a clay or shale brick is 0.0000036 inches per inch per °F. For each 100 linear feet of brick wall for a 100°F temperature rise, it will expand 0.43 inch or about 7/16 inch.

One of the principal causes of masonry cracking is differential movement between masonry and the various elements of the structure with which the ma-

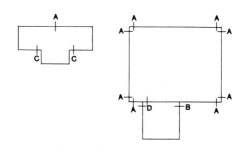

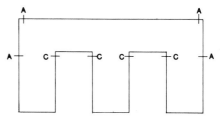

FIGURE 4-8 Typical expansion joint placement. Courtesy Brick Institute of America.

sonry abuts. The skeleton-frame, both steel and reinforced concrete move at different rates than the masonry as a result of differing coefficients of thermal expansion and contraction. Differential thermal movement also occurs between masonry walls and concrete foundations, between parapet walls and enclosing walls, and at offsets in walls and intersecting walls. Each building design must be analyzed to determine potential movements and provision made to relieve excessive stress by the introduction of expansion joints.

The Brick Institute of America (BIA) suggests typical expansion joint placement in walls as shown in Figure 4-8.

Details for these typical expansion joints may be found in BIA Technical Notes—Series 18A and 18B.

While the above differential thermal movements occur horizontally and require vertical expansion-relieving joints, there is also a masonry failure that is induced by the shortening of structural frames that requires the use of horizontal relieving joints. In high-rise buildings, plastic flow, especially in lightweight concrete columns, may cause a vertical shortening of the frame. Since the exterior masonry is usually carried on shelf angles, there is the possibility of masonry failures characterized by bowing, or by spalling of the faces at shelf angles and lintels. The introduction of horizontal expansion joints at these angles is critical; BIA Technical Note 18A recommends a detail for this condition.

REINFORCED BRICK MASONRY

Reinforced brick masonry (RBM) is a system consisting of brick, mortar, grout, and reinforcing steel, designed and placed in such a manner that resultant masonry construction will have greatly increased resistance to forces which produce tensile, shearing, and compression stresses. Natural phenomena which produce these forces are earthquakes, and strong winds such as hurricanes, tornadoes, and cyclones. RBM may also be used for isolated elements of a building, such as chimneys, beams, girders, piers, floors, roofs, and lintels.

The principles used in the design of RBM are essentially the same as those used and commonly accepted for reinforced concrete; therefore, similar formulae may be used. Standards used in the design of RBM may be found in American National Building Code Requirements for Reinforced Masonry, A41.2, and in the Brick Institute of America "Technical Notes" Series 17. Mortar and grout used in the RBM system are governed by ASTM C476.

In the 1950s a method of construction known as the "High Lift Grouting System" was developed in the San Francisco Bay Area. This technique has improved over the years and evolved into a more sophisticated and economical solution in its application to both low- and high-rise structures using RBM. In this system the outer wythes of masonry are built up around the vertical reinforcement and allowed to set for a minimum period of 3 days; after that, the grout is pumped into the space containing the reinforcement. Figure 4-9 illustrates a typical wall constructed by the high-lift grouting technique.

PREFABRICATED BRICK MASONRY

Prefabricated brick masonry is a relatively new masonry system that has certain advantages over conventional in-place brick laying.

Developed in Europe in the 1950s, the system was adapted in the United States in the early 1960s. Begun initially to mechanize the panelization of brick to overcome cold weather construction, it has grown in scope to include the fabrication of complex shapes.

The advantages of prefabrication include:

1. Elimination of scaffolding
2. Fabrication of complex shapes
3. Factory prefabrication year-round under controlled weather conditions
4. Elimination of enclosing and winterizing the structure under construction
5. Reduction in construction time.

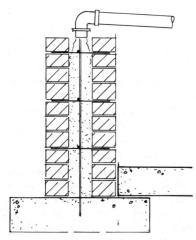

FIGURE 4-9 High-lift grouting for RBM wall. Courtesy Brick Institute of America.

FIGURE 4-10 RBM panel, 39 feet long, being hoisted into place. Courtesy Brick Institute of America. This illustration also appears in the color section.

Prefabricated brick masonry has been made possible by the development of high-bond mortars, a rational design method for brick masonry, and improved brick units. In addition, a standard for prefabricated panels has been established in ASTM C901, Specifications for Prefabricated Masonry Panels.

Prefabricated panels are also possible with thin brick facing units in conjunction with concrete, fiberboard, or other backing materials. Figure 4-10 shows a representative example of prefabricated masonry element being hoisted into position.

CONCRETE MASONRY UNITS

HISTORY

Modern-day concrete block or concrete masonry unit (CMU) dates back to 1882. Today, with increased automation and the benefit of long-term use, experience, and research, concrete masonry units have become not only a standard structural building material but also a product having aesthetic architectural applications. This latter quality has been achieved through the incorporation of color, texture, surface treatment, and shapes of various configurations.

MANUFACTURE

Concrete masonry units are manufactured using portland cement, graded aggregates, and water. When required, the mix may also contain air-entraining admixtures, coloring pigments, and pozzolanic materials. The manufacturing process utilizes machine molding of a rather dry, no-slump concrete. Automatic machinery consolidates and molds thousands of units each day.

Curing, which is absolutely essential to control shrinkage, is performed in a number of ways. Air drying, which requires about 2 to 4 weeks of open-air drying with weather protection, is not as desirable as two types of accelerated curing. The more prevalent type of accelerated curing requires heating the units in a steam kiln at atmospheric pressure, with steam temperature at about 120 to 180°F for about 12 to 18 hours. The other accelerated method is autoclaving or high-pressure steam curing. This latter method subjects the units to saturated steam at about 375°F and a pressure of 170 psig for about 12 hours.

Types

There are two major concrete masonry blocks and a concrete brick manufactured to meet ASTM requirements as follows:

ASTM C90—Hollow Load-Bearing Concrete Masonry Units
ASTM C145—Solid Load-Bearing Concrete Masonry Units
ASTM C55—Concrete Building Brick

A hollow unit is defined as having a net concrete area of less than 75%, and a solid unit as having a net concrete area of more than 75%.

TABLE 4-9
GRADE USES OF CONCRETE MASONRY UNITS

Grade	Block	Brick
N	For general use as in exterior walls below and above grade that may or may not be exposed to moisture penetration or the weather and for interior walls and back-up.	For use as architectural veneer and facing units in exterior walls and for use where high strength and resistance to moisture penetration and severe frost action are desired.
S	Limited to use above grade in exterior walls with weather protective coatings and in walls not exposed to the weather.	For general use where moderate strength and resistance to frost action and moisture penetration are required.

TABLE 4-10
MOISTURE CONTENT REQUIREMENTS

% Linear Shrinkage	% Moisture Content		
	Humid	Intermediate	Arid
0.03 or Less	45	40	35
0.03 to 0.045	40	35	30
0.045 to 0.065	35	30	25

Humid—average annual RH above 75%. Intermediate—average annual RH 50 to 75%. Arid—average annual RH less than 50%.

The two blocks and the concrete brick have two grades, N and S, as per ASTM standards which establish their conditions of use (shown in Table 4-9).

Since concrete masonry units are subject to shrinkage and expansion, for architectural purposes it is best to confine selection and use of these products to the types described hereinbefore. Also limit use to the moisture controlled units, designated as Type 1 under the ASTM standards and as shown in Table 4-10.

Aggregate Selections and Block Weight

Aggregate selection has an influence not only on the weight of the block but also on the textural quality, including the split-face units. There are three weight gradations of CMU, as shown in Table 4-11.

TABLE 4-11
BLOCK WEIGHTS AND AGGREGATES

Weight Designation	Unit Weight (pcf)	Aggregates Types
Normal weight	Over 125	Sand and gravel Crushed stone and sand
Medium weight	105–125	Air-cooled slag
Light weight	Less than 105	Coal cinder Expanded slag Expanded clay, shale, slate Pumice Cellular concrete

Sizes and Shapes

Concrete block dimensions are usually given in nominal sizes, with the face size normally 8 x 16 inches. Since joint dimensions are typically ⅜ inch thick, the actual face size is 7⅝ x 15⅝ inches so that in-the-wall dimensions add up to 8 x 16 inches. Block is commonly available in thickness (width) of 2, 4, 6, 8, 10, and 12 inches. The most common nominal block face is 8 x 16 inches.

Concrete brick is likewise sized to be laid with a ⅜-inch joint, resulting in modules of 4-inch widths and 8-inch lengths. The typical concrete brick is 2¼ x 3⅝ x 7⅝ inches.

SURFACE FINISHES

For architectural purposes, concrete block is now being produced with a variety of textures which are influenced by the type and grading of aggregates; with color by the addition of various pigments; and with prefaced units having surface-applied resinous materials and ceramic or porcelainized glazes in various colors.

Textured Surfaces

The surface textures available for concrete masonry units may be classified as fine, medium, and coarse and are achieved by various aggregates, their grading, the mix proportions, and the wetness of the mix. For large projects some manufacturers may produce special textures to suit the architect's needs. See Figure 4-11, Surface Texture.

A smooth-faced block is available that is produced by grinding the exposed faces approximately ⅟₁₆ to ⅛ inch. Ground-face units show aggregates of varying colors, sizes, and types to good advantage. With the addition of coloring pigments, these ground-face units are used in walls and partitions as a finishing material without further treatment.

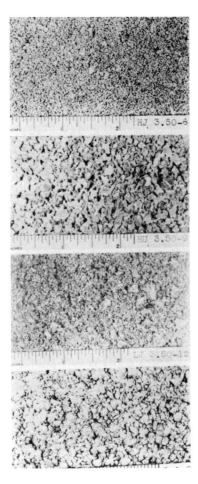

FIGURE 4-11 Examples of textures of concrete masonry units. (Used with permission of Portland Cement Association.)

Colors

The natural color of concrete masonry depends upon the color of the aggregates and the cement used. This can vary from the typical shades of gray to the earth tones of brown. Since these colors are achieved by the use of the locally available materials, the resulting colors will be more uniform and easier to duplicate in the event of additional work at the same project. Mock-ups of such units should be erected to ascertain the maximum degree of variation that will be permitted.

In addition to the natural color variations available, artificial colors can be obtained through the addition of mineral oxide pigments in the concrete mix. However, when used on the exterior, the colorfastness of

some mineral oxides such as blue are questionable. Another problem associated with the exterior use of some of these units is the possibility of efflorescence marring their appearance.

Prefaced Units

Concrete block may also be prefaced with a resinous mixture of resin, sand or portland cement with integral colors, to produce smooth or dappled surfaces approximately ¼ inch thick. The resinous facings form an integral bond with the concrete masonry units and are cured by a heat process.

The facings are about ¹⁄₁₆ inch longer than the block face and return ¼ inch into the joint. The resulting bed and head are ⅜ inch thick, but the face joint becomes ¼ inch thick. Double-faced units are also available along with such features as caps, sanitary bases, returns, and scoring. The resinous facings are produced to meet the requirements of ASTM C744.

Split-Face Units

Block and bricks may be split or fractured to produce a rough finish. The appearance of the face when split will be irregular, and the aggregate within the face will be fractured to provide interesting variations of texture. By varying the cement aggregates and their gradation and by adding color pigments, a host of possibilities is available for architectural expression.

Sculptured and Screen Units

Sculptured or patterned block are produced by providing raised or recessed geometric forms cast into the faces, thus providing the architect with a wide palette from which to select. In addition, patterns and profiles may include, ribbing, fluting and scoring, all as shown in Figure 4-12. The sculptured surfaces provide a display of striking shadow effects which can be varied by the specific pattern in which the units are laid.

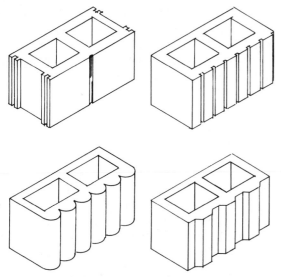

FIGURE 4-12 Ribbed, fluted and scored block. (Used with permission of Portland Cement Association.)

Screen or pierced units are used for a variety of purposes such as solar screens, wind breaks, diffusion of strong sunlight, and decorative dividers. They offer privacy with a view and airy comfort with wind control. However some designs are only available in certain areas and some designs are restricted by copyright. Construction of walls and partitions using these units requires specific structural analysis; the use of a framing system; vertical and horizontal joint reinforcement; and anchorage of the wall to a framing system. See Figure 4-13, Screen Walls.

FIGURE 4-13 Screen walls provide privacy. (Used with permission of Portland Cement Association.)

FIGURE 4-14 Acoustical units employed to improve acoustics. (Used with permission of Portland Cement Association.)

Acoustical Units

Special units providing sound-absorbing qualities are produced with vertical-slotted faces. The sound absorption is derived from the fact that the cavity within the unit acts as a sound resonator. Other units have the cavity filled with fibrous material or an energy-absorbing septum. Each type offers sound absorption at differing frequencies and is useful in reducing noise in such applications as gymnasiums, natatoriums, and industrial applications. See Figure 4-14 for an application involving acoustical units.

BLOCKWORK

Masonry unit walls and partitions are usually constructed as one wythe thick. However, when exterior walls are designed in locations where rain penetration may be a factor it would be prudent to consider a cavity wall design. See Wall Design, Weathertightness, in this Chapter.

Units delivered and stored at the job site should be protected from rainfall or moisture gain in humid areas by canvas or polyethylene tarpaulins. During construction it is essential that the tops of walls be covered with tarpaulins to prevent entry of rain or snow. Concrete masonry units should never be wetted before setting, since moist block will shrink as the excess moisture evaporates. It is essential that the block be stored and erected at the moisture content required in the specifications.

Mortars

Mortar for concrete block masonry is similar in most respects to that used for brickwork as described in this chapter. When erecting composite walls of brick and block, the mortar specified for the brick is used throughout.

When colored mortar is desired, the desired results may be achieved by the use of white cement and pigments, colored masonry cement, or colored sand.

Joints

Mortar joints are normally ⅜ inch thick. Two types of mortar beds are used: full bedding (webs and face shells are bedded in the mortar) and face-shell bedding (only the face shells are bedded in mortar). Full bedding is used where strength is an important factor (i.e., first course on a foundation, solid units, piers, columns, and pilasters) and in reinforced masonry. For all other concrete block, face-shell bedding is sufficient. See Figure 4-15 for types of mortar bedding.

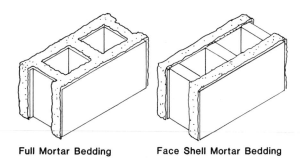

Full Mortar Bedding   Face Shell Mortar Bedding

FIGURE 4-15 Types of mortar bedding on block. (Used with permission of Portland Cement Association.)

To obtain more watertight construction, a concave joint or a Vee joint should be used. (See Figure 4-3, mortar joints.) Concave and Vee joints compress the mortar and offer better resistance to water penetration than other types of joints.

Crack Control

While a number of factors will cause concrete masonry walls to move (e.g., thermal change and change in moisture content), shrinkage due to moisture loss is quite significant and should be addressed by the designer to introduce measures to inhibit crack formation.

Four factors influence the degree of drying shrinkage: (1) type of aggregate; (2) curing method; (3) method of storage; and (4) protection of freshly set walls from rain and dew. Units made with sand and gravel shrink less than other types. High-pressure steam curing reduces shrinkage best. Moisture-con-

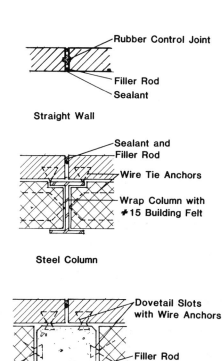

Rubber Control Joint
Filler Rod
Sealant

Straight Wall

Sealant and
Filler Rod
Wire Tie Anchors
Wrap Column with
#15 Building Felt

Steel Column

Dovetail Slots
with Wire Anchors
Filler Rod
and Sealant

Concrete Column

FIGURE 4-16 Control joints in block.

trolled units when installed at or below the relative humidity of the locality will exhibit less shrinkage. Insofar as moisture content is the controlling factor affecting shrinkage, masonry units should be stored and kept dry until setting.

Two methods can be utilized to accommodate shrinkage. One is to introduce control joints, the other is to provide joint reinforcement. Figure 4-16 shows typical control joints.

Location of control joints as recommended by ACI Committee 531 "Concrete Masonry Structures Design and Construction" are shown in Table 4-12.

For more detailed descriptions of locations of control joints see the *Concrete Masonry Handbook for Architects, Engineers & Builders* published by the Portland Cement Association.

Joint reinforcement is utilized to control cracking caused by drying shrinkage. While reinforcement does not eliminate cracks, it controls the formation of highly visible cracks. The stresses which do occur are transferred to the reinforcement; as a consequence there is an even distribution of stresses, and hair line cracks result which are barely visible.

Joint reinforcement should consist of a minimum of 9 gage side rods and 9 gage cross rods, complying with ASTM A82 and located as follows:

1. In the first two courses above and below all openings, and extending a minimum of 2 feet on both sides of opening.
2. In the first two or three joints above floor level, below roof level and near top of the wall.
3. As shown in Table 4-12.

TABLE 4-12
MAXIMUM SPACING OF CONTROL JOINTS IN NON-REINFORCED MASONRY

Maximum Spacing of Joint Reinforcement (inches)	Maximum Spacing of Control Joints	
	Panel Length/Height	Panel Length (feet)
None	2	40
24	2.5	45
16	3	50
8	4	60

For additional information on sizes, galvanizing, and locations of joint reinforcement, review the current literature of the manufacturer of these materials.

Bond Patterns

Exposed concrete block can be used quite effectively architecturally by creating numerous bond patterns to achieve various effects. Long, low expressions may be obtained with the use of 2-inch high units laid horizontally. Height may be expressed by vertical stacking of 8 x 16-inch units. Sculptured and patterned block can create rhythmic sequences. The variety of bond patterns that can be achieved depends only upon the imagination of the architect.

Ties

Bonding of concrete masonry units in composite wall construction is best accomplished with metal accessories. While masonry bonding is possible, there are certain risks. Rain penetration is more likely with masonry bonds than with metal ties. Metal ties also allow the flexibility of differential movement when composite construction is used, thereby relieving stresses and limiting cracking.

Ties for exterior use are usually fabricated of galvanized steel, stainless steel, or other types of corrosion-resistant rods or wire. For cavity walls, a crimp is introduced in the wire that is centered over the cavity to act as a drip. For composite walls of block and brick facing, an adjustable tie system is used that allows for adjustment of the difference in levels between the wythes. Continuous metal ties are more commonly referred to as joint reinforcement. See Figure 4-17 for representative metal ties.

WALL DESIGN

Concrete walls and enclosures possess certain physical properties, as well as the aesthetic. They have the ability to handle thermal problems, the ravages of fire,

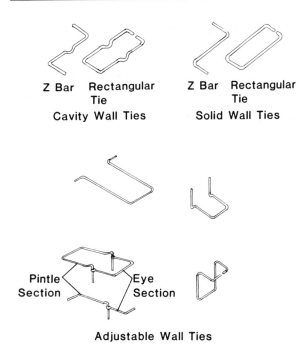

Z Bar   Rectangular   Z Bar   Rectangular
          Tie   Tie
    Cavity Wall Ties   Solid Wall Ties

Pintle   Eye
Section   Section

Adjustable Wall Ties

FIGURE 4-17 Concrete unit masonry ties. (Used with permission of Portland Cement Association.)

noise, and rain penetration, all if properly addressed and designed.

Thermal Characteristics

While masonry is a poor insulator of heat, the addition of insulation in a cavity wall, or the use of insulation in the cores will greatly enhance the thermal efficiency of a concrete masonry wall. Table 4-13 illustrates the U-values of various walls.

TABLE 4-13
U-VALUE FOR VARIOUS MASONRY WALLS

Wall Construction	Insulation	U-Value
8 in. sand and gravel	none	0.53
" "	loose fill-in core	0.35
8 in. expanded clay, shale	none	0.33
" "	loose fill-in core	0.18
10 in. cavity wall, 4 in. heavyweight block, 4 in. brick	none	0.34
" "	poured fill-in cavity	0.13

Fire Resistance

Fire-resistant assemblies are tested in accordance with ASTM E119, Fire Tests of Building Construction and Materials. Fire-resistive ratings of concrete block depend upon the type of aggregate used and the *equivalent thickness* (ET) of the block, where the ET is the average thickness of solid material. Blocks with pumice aggregate provide the highest hourly ratings.

Most buildings codes will state the required fire resistance of concrete masonry units in terms of the ET.

Sound Insulation

Sound transmission (the passage of sound from one space to another) and sound absorption (the property of a material or construction to absorb sound energy) are factors to be considered in designing intervening walls and partitions.

Exterior walls at airports, major highways, and similar noise-generating sources can be designed with masonry walls that will effectively reduce the sound transmission. Table 4-14 shows some masonry constructions and their sound transmission loss values.

For interior partitions where sound absorption is desirable, as in gymnasiums, swimming pool areas, and auditoriums, the use of acoustical block will provide increased sound absorption characteristics. Lightweight aggregate block will have noise reduction coefficients (NRCs) in the range of 0.40 to 0.45, whereas some acoustical blocks with slotted faces and fibrous fillers have NRCs of 0.60 to 0.70. However, when paint is applied to CMU the NRC will be reduced considerably.

Weathertightness

Rain can penetrate blockwork between mortar joints and the masonry units, or through shrinkage cracks in the block. In solid composite concrete block wall construction it is essential to back parge the vertical

TABLE 4-14
SOUND TRANSMISSION LOSS OF CMU WALLS

Construction	STC Value
8-in. hollow, lightweight aggregate	49
8-in. hollow, stone aggregate	52
10-in. cavity, 4-in. brick, 4-in. lightweight hollow units	54
8-in. hollow, lightweight aggregate with ½-in gypsum board on resilient channels	56
10-in. cavity, 4-in. brick, 4-in. lightweight aggregate with ½-in.gypsum board on resilient channels	59

Note how cavity-type walls enhance the STC value and how the addition of resilient channels and ½-in. gypsum board likewise reduces the sound transmission.

collar joint between the interior and exterior wythes, including filling of all mortar joints. However, a well-designed and constructed cavity wall will outperform a solid wall in the resistance to rain penetration. As noted previously, concave or Vee joints are best suited to resist water penetration.

Paints, coatings, and stucco are sometimes used to reduce water penetration of uncoated block walls. To determine the effectiveness of these applications, a test procedure for ascertaining the resistance to leakage of unit masonry walls subjected to wind-driven rain may be utilized. ASTM E514 "Water Permeance of Masonry" may be used to measure the appearance of moisture or visible water on the back of a panel that has been coated. The degree to which the coating prevents or permits the passage of water under these test conditions will indicate its effectiveness as a waterproofer.

REINFORCED CONCRETE MASONRY

Reinforced concrete masonry, like reinforced brick masonry, consists of block, mortar, reinforcing steel, and grout so designed and constructed that the assemblage works in unison to resist forces associated with seismic conditions and heavy winds.

Units are laid so that unobstructed vertical cells are aligned to form a continuous core in which reinforcing steel and grout are placed. Mortar and grout used in connection with reinforced concrete masonry are specified in ASTM C476.

Two techniques are used to erect reinforced concrete masonry: (1) low-lift grouting and (2) high-lift grouting. The low-lift grouting procedure is much more simple and more widely used. In low-lift grouting for a single wythe wall, the masonry units are laid to a height of 5 feet and the grout poured into the cores embedding the steel reinforcement. In high-lift grouting, the units are laid story high and then filled with grout by pouring or pumping.

Standards used in the design of reinforced concrete masonry are contained in American National Standards Building Code Requirements for Reinforced Masonry, A41.2; additional data are available from the National Concrete Masonry Association and the Portland Cement Association.

PREFABRICATED CONCRETE MASONRY PANELS

Masonry unit panel walls are being fabricated both on-site and off-site by a variety of methods. More sophisticated developments include the use of machines that lay block and fill the head, face, and cross-web joints with mortar.

The advantages of this mechanized production are: (1) an all-weather operation; (2) quality control and uniformity of production; and (3) speed of erection.

MORTARLESS CONSTRUCTION

A recent innovation in concrete block construction is the development of a unit that has a system of tongues and grooves at each end and at top and bottom. This system of tongues and grooves is designed so that the block can be laid in a wall without mortar, and by interlocking they fit tight without shifting.

Blocks are manufactured with the various special units required for corner block, bond beams, jamb units, and caps. Face sizes are 8 x 16 inches, which is the normal equivalent of block and mortar joints.

Test data for fire resistance, loading, and water penetration resistance, are available from the developer of the product, Mc IBS, Inc., St. Louis, Missouri.

STRUCTURAL GLAZED FACING TILE

Structural glazed facing tiles are made from clay, shale, fire-clay, or mixtures thereof. They have a finish consisting of a ceramic glaze fused to the body at above 1500°F. A wide choice of colors is available.

The clay tile body is produced to meet ASTM C212 "Structural Clay Facing Tile" for a load-bearing tile. The ceramic glazed facing is produced to comply with ASTM C126 "Ceramic Glazed Structural Clay Facing Tile." Two grades are available: Grade S (select) for use with comparatively narrow joints; and Grade SS (select size or ground edge) for use where variation of face dimension must be very small. Units are also available as single-faced or two-faced.

Two face sizes are produced: a nominal 5⅓ x 12 inches and a nominal 8 x 16 inches, with bed depths of 2, 4, 6, and 8 inches.

Acoustical units having perforated faces and with NRCs of 55 to 50 are produced for use in such locations as indoor swimming pools, gymnasiums, and laboratories.

Units may be laid with joints as thin as ³/₁₆ inch, either in stretcher bond or stack bond.

ARCHITECTURAL TERRA COTTA

Architectural terra cotta was a widely used material up until the late 1920s. A striking example of its use is the facade of the Woolworth building in New York City.

Architectural terra cotta is generally custom made, hard burned, glazed or unglazed clay building units, plain or highly ornamental, and hand molded or machine extruded. The units are hollow or open back with ribs, finished in a variety of glazes and complex shapes and designed primarily for ornamentation and decoration. A more modern version is ceramic veneer, an architectural terra cotta characterized by large face dimensions and thin sections.

Architectural terra cotta is anchored to backing materials with metal ties. Ceramic veneer is adhered to backing with mortar. More recent developments in the use of precast panels incorporates the use of key-backs, which permits concrete to flow into the key-back and to create a mechanical bond.

GLASS BLOCK

Glass block was first introduced as a building material in the early 1930s and its use has been somewhat erratic. Overused inappropriately in some cases, its use declined. In recent years, however, physical changes in styles, colors, sizes, and light transmission, and more innovative use by architects have stimulated the reuse of this building material.

Glass block is generally produced as a hollow unit by fusing two halves together and creating a partial vacuum during the process. Varying the pattern by pressing into either the inner or outer surfaces of the block face will cause light striking the surface to be diffused, reduced, or reflected differently. Solid units are also available.

Fibrous glass inserts can be installed within the unit to control glare, brightness, and solar heat gain and to increase thermal transmittance. Solar reflective units have a highly reflective, thermally bonded, oxide surface coating that reduces both solar heat gain and transmitted light.

Units are generally square, nominally 6, 8, and 12 inches, with some rectangular sizes 4 x 8 inches and 6 x 8 inches. Bedding widths are 3⅛ inches and 3⅞ inches. For styles, sizes, and patterns, it is best to consult the most recent manufacturer's literature. Since the units cannot carry more than their own weight wall panels are restricted as to length, height, and square foot area. The current literature of manufacturers should be consulted with respect to these limitations, and also for precautions that should be taken to accommodate expansion at heads, jambs, and sills.

Glass block is laid in stack bond with ¼-inch mortar joints. Joints must be reinforced as recommended by manufacturers and as required by building codes.

STONE

GENERAL

Rock Classification

Rock is a geological term referring to the solid material that forms the earth's crust and is composed of an aggregate of grains of one or more minerals. Dimensional stone is rock selected for building use and processed by shaping, cutting, or sizing. It is also the commercial term applied to quarry products.

Rock or stone is divided into three classifications based on its geologic origin: igneous, sedimentary, and metamorphic.

Igneous rock is formed from the solidification of molten rock such as that caused by volcanic activity and pressure caused by the shifting of the earth's surface. The most widely used igneous rock used as a building stone is granite.

Sedimentary rock is formed from silt, marine life, and disintegrated rock that has been deposited in place or as sediment by running water in rivers or seas. These deposits have been layered and solidified through pressure induced by overlying materials and by cementing together through chemical action such as gases contained in water. Some examples of building stones derived from sedimentary rock are limestone, sandstone, and travertine.

Metamorphic rock may be either igneous or sedimentary rock whose character and structure have been further changed by subsequent pressure, heat, moisture, or combinations of these forces. Examples of metamorphic rock are marble and slate.

Building Stone

Rock used in building construction is known as building stone. Its use encompasses many aspects of a building: (1) as a structural material in load-bearing walls; (2) as a finish material for exterior and interior walls; and (3) as a flooring and paving material.

With new techniques for quarrying, and with new equipment for finishing, the search for new sources of stone to provide a wider range of colors has ensued. Some stone is available in limited quantities and in limited areas, while others are plentiful worldwide.

Whenever domestic or foreign stone is considered for use on the exterior of a building, care must be exercised in its selection. Stones indigenous to the locality where they are quarried have generally withstood the ravages of their environment and may be used in similar climatic environments without trepidation. However, occasionally when stone is used in a new locale without proper investigation, unforeseen problems arise. The brownstone materials used for exterior steps and facades of residences in New York City deteriorated badly. The ravages of water, wear, and temperature that the brownstone had to endure in New York were completely different from those conditions prevalent at the quarries from whence these stones were taken. One would not use adobe brick in a wet climate, but it endures in the arid Southwest.

Similarly, building stone that has endured on structures through the ages in its native environment is beginning to show the effects of acid rain caused by industrial fumes and automobile engine exhaust gases, a problem that the designers selecting the building stone could not envisage at that time. The chief culprits in the disintegration of stone are CO_2 and SO_4, the important deleterious ingredients found in the combustion of fossil fuels.

Stone Masonry Set in Mortar

Stone masonry construction, as noted previously, has undergone dramatic changes as architectural styles have changed and the curtain wall stone panel or veneer panel has come into vogue. The traditional stone masonry wall set in mortar and used for fine residences, religious buildings, public buildings, banks, and other load-bearing walls is classified as to shape and finish as being either rubble or cut stone. Rubble is essentially the crude, uncut stone that is collected as fieldstone having rounded, natural faces or angular, broken faces. When laid in a wall the pattern can be random or coursed. See Figure 4-18 for fieldstone elevations.

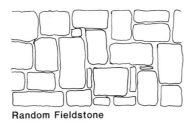

Random Fieldstone

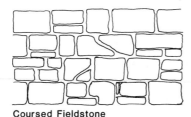

Coursed Fieldstone

FIGURE 4-18   Rubble or fieldstone coursing.

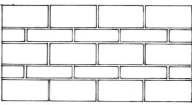

Regular Coursed Ashlar

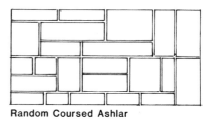

Random Coursed Ashlar

FIGURE 4-19   Ashlar wall construction.

For cut stone, there are essentially no standard definitions. The terms cut stone, ashlar, and dimension stone have been used with varying degrees of differences among producers of marble, granite, and limestone. The term ashlar refers both to (1) a stone that has been cut generally square or rectangular and with either a finished or unfinished face and (2) a stone masonry wall construction utilizing ashlar stone. See Figure 4-19 for ashlar construction.

ASTM has but one definition for dimension stone: "a natural building stone that has been selected, trimmed or cut to specified or indicated shapes or sizes, with or without one or more mechanically dressed surfaces."

Bonding of Stone Masonry

Where stone masonry is laid up with a backing of either stone, concrete block, or brick, the bond between the facing and the back-up may be accomplished by use of bond stones, anchors, bond courses, or a combination of these elements.

Bondstones are facing stones generally cut to twice the bed thickness of the material used for facing stones. The percentage of stone used as bondstones is usually defined by building codes. Figure 4-20 shows a typical wall bonded with bond stones.

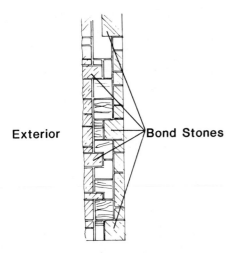

Exterior ← → Bond Stones

FIGURE 4-20 Bonding of stone masonry.

Veneer Stone

With the advent of new construction techniques and architectural styles, a change has occurred in the manner in which building stone is used on the facades of high-rise structures. The introduction of skeleton frame construction has almost eliminated the need for thick, massive, load-bearing stone masonry walls. The trend has been toward building stone panels or veneers less than 2 inches thick (50 mm), for granite, marble, travertine, and slate. To accomplish these ends, special anchoring systems have been developed to support these thin veneers. In some instances veneers of granite, marble, travertine, and slate measuring 40 mm (1.57 in.) and 30 mm (1.38 in.) have been utilized for exterior facings using standard and newly patented anchoring systems.

The Zibell Anchoring System is a patented product of the Georgia Marble Company. The system includes galvanized steel struts that can be anchored to either concrete or steel back-up and extruded aluminum members and aluminum fastening clips, all of which form a metal grid that carries the stone veneer facade. It is also adaptable to renovations and interiors. The system lends itself to simple and rapid installation.

The ISR System is a development of the Vermont Marble Company. This system is designed for rapid installation of prefabricated panels of marble, granite, serpentine, or other natural building stones from 7/8 inch to 1½ inches thick. Stainless steel angles are bolted to a back-up structural grid or masonry wall.

The angles support the stone veneer panels by engaging grooves at the edges of the panel.

The architect or designer must investigate the thickness of the particular species of stone that is under consideration with respect to face size, modulus of rupture, flexural strength, compressive strength, and handling stresses to ensure that the thickness is structurally adequate for the intended use.

With the introduction of thin veneers there has also been a reduction in the joint width, and sealants have replaced mortar in the joints. This replacement of mortar by sealants has occurred because narrower mortar joints have a tendency to powder out. Sealant technology has improved to the point where sealants provide a jointing material far superior to mortar. To assure non-staining of stone by sealants, use test method ASTM C510.

Typical Building Stones and Uses

While exotic stone such as Mexican onyx and other rare stones have been used in building construction, this book is concerned with the most commonly used building stones. Table 4-15 lists these stones, their characteristics, and areas of use.

Physical Properties

Since building stone is one of the earliest building materials, a wealth of information has been accumulated on its physical and chemical properties. Building stone was used early on as a structural material in wall-bearing construction, and a good deal of data are available on thicknesses, sizes, and details associated with this type of construction. However, since World War II, more and more veneer-type panel construction has been used. Designers must therefore be extremely careful in selecting stone species, thicknesses, and panel sizes before making decisions on these characteristics. The factors of weathering, wind loads, and seismic conditions where these prevail must be taken into account since there is not yet a sufficient long-term history of thin, veneer-type building stones under these conditions of use.

There are a number of ASTM test methods available to ascertain certain physical characteristics that are

TABLE 4-15
TYPICAL BUILDING STONES AND USES

Building Stone	Type	Color	Major Use	Minor Use
Granite	Igneous	Wide range	Exterior and interior wall facings	Paving flooring
Limestone	Sedimentary	Buff, gray	Exterior wall facings	Copings, sills, interior wall facings
Marble	Metamorphic	Wide range	Exterior and interior wall facings, flooring	Countertops
Sandstone	Sedimentary	Yellow, brown, reds, tan	Exterior wall facings	Paving
Slate	Metamorphic	Blue, gray, green, red, black	Paving, roof shingles	Wall facing
Travertine	Sedimentary	Tan, buff, gray	Exterior and interior wall facings	Flooring, paving

pertinent when designing with building stone as follows:

ASTM C241	Abrasion Resistance
ASTM C170	Compressive Strength
ASTM C97	Density
ASTM C880	Flexural Strength
ASTM C99	Modulus of Rupture
ASTM C510	Staining of Stone by Sealants
ASTM C97	Water Absorption
ASTM C217	Weather Resistance of Slate

Structurally, building stones used for building construction are amply strong, especially in compression. When used as a soffit or ceiling, the thickness may have to be increased to accommodate anchors, which reduce the section.

Durability (see Chapter 1) is a most important performance characteristic, and weathering is the chief factor. Local stones may be used in the same locale if a history of their use there shows successful applications over a period of time. In addition, durability may be judged by examining open quarries or outcroppings to see how the stone has fared when exposed to the elements over a long time. That specific type of stone may also be expected to perform well in similar climatic environments far removed from its original source. However, as has been stated previously, industrial, chemical, and automobile engine exhaust fumes may reduce the life expectancy. If a public building or a monument is contemplated with a life expectancy of several hundred years, then a more searching examination should be conducted to determine durability. If the structure is to have a life expectancy of 75 to 100 years, the risks of deterioration are much less.

For interior walls, most building stones, including the exotic marbles, will have no durability problems. However, when used as countertops, or when used in toilet rooms as walls, wainscots, or partitions, there may be evidence of staining or discoloration. Proper maintenance or the use of protective sealers may inhibit staining.

Table 4-16 shows some of the physical characteristics of typical stones used in building construction.

The National Building Granite Quarries Association recommends for granite the following standards for physical characteristics:

Water Absorption	0.4%	max.	ASTM C97
Density	160	lbs/cu ft	ASTM C97
Compressive Strength	19000	psi min.	ASTM C170
Modulus of Rupture	1500	psi min.	ASTM C99

TABLE 4-16
PHYSICAL CHARACTERISTICS OF TYPICAL BUILDING STONES

Stone	Compressive Strength	Modulus of Rupture	Absorption
Granite	20,000–36,000 psi	1480–3240 psi	0.08–0.22
Limestone	4000–20,000 psi	700 psi min.	7.5% max.
Marble	6000–17,000 psi	—	0.069–0.609
Sandstone	4000–64,000 psi	1500 psi	—
Slate	10,000–15,000 psi	12,000 psi	.01–.02%

The Indiana Limestone Institute recommends these values when designing with oolitic limestone:

Water Absorption	7½% max.	ASTM C97
Modulus of Rupture	700 psi max.	ASTM C99
Compressive Strength	4000 psi min.	ASTM C170

ASTM has developed a set of standards for the major building stones that set forth the requisite working characteristics as follows:

Physical Property	Test Req.	ASTM Test
Granite—ASTM C615, Architectural Grade		
Absorption	0.4%	C97
Compressive Strength	19000 psi	C170
Modulus of Rupture	1500 psi	C99
Limestone—ASTM C563, Medium Density		
Absorption	7.5%	C97
Compressive Strength	4,000 psi	C170
Modulus of Rupture	500 psi	C99
Marble—ASTM C503		
Absorption	0.75%	C97
Compressive Strength	7500 psi	C170
Modulus of Rupture	1000 psi	C99
Sandstone—ASTM C616, Quartzitic		
Absorption	3%	C97
Compressive Strength	10,000 psi	C170
Modulus of Rupture	1000 psi	C99
Slate—ASTM C629		
Absorption	0.25%	C97
Modulus of Rupture	9000 psi	C120

In some instances, particularly for nondomestic stone, some physical characteristics may fall below the working criteria established by the ASTM standards. The prospective user may be enthralled by the exotic stone being investigated and must decide whether the deficient characteristic is sufficient to rule out the choice of material or whether the deficiency can be compensated for by some other means, such as increased thickness.

Anchors and Anchorage

Anchors are metal ties used to secure stone in place. Anchors include straps, rods, dovetails, expansion bolts, cramps, and wire ties. Most anchors are designed to hold stone in its vertical position rather than support its gravity loads.

Under the heading of thin veneer stone, examples of proprietary and some standard metal anchoring systems for thin veneers are discussed. Stone exceeding 2 inches in thickness generally requires the types of anchors shown in Figure 4-21.

The metals used for anchors should always be noncorrosive, (1) so that they do not fail by virtue of corrosion and (2) so that they do not stain the stone they are securing. Stainless steel for anchors offers many attributes. It is not too costly. It has high tensile strength. It is practically noncorrosive and most enduring. Other metals used are zinc alloys, yellow brass, and commercial bronze, with the least desirable being galvanized steel.

Whenever metal anchoring systems are used to support stone, care should be exercised to avoid galvanic action in unlike metal connections. See Chapter 1, Compatibility.

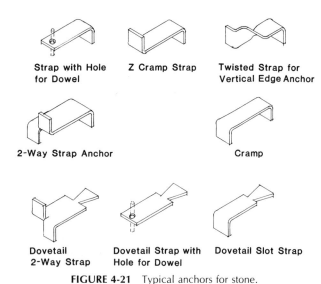

FIGURE 4-21 Typical anchors for stone.

Staining

Building stone does have a tendency to absorb some moisture. When this moisture migrates from back-up materials, anchorage systems, and from ground-level sources, the moisture entering the stone may carry contaminants or soluble salts that may stain the stone. A major source of soluble salts is the concrete or masonry back-up.

To prevent staining, a number of precautions must be observed. By introducing weep holes above shelf angles and at base courses, the accumulation of water is reduced. Dampproofing treatments provided at base courses will act as a barrier to moisture migration into the stone. See Figure 4-22.

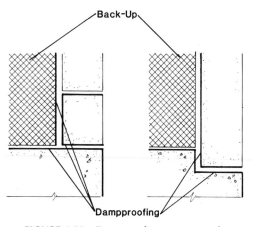

FIGURE 4-22 Dampproofing stone at grade.

The best defense against staining is to prevent moisture from entering the wall. The most vulnerable areas are the stone joints. In veneer work, the use of sealant joints in lieu of mortar joints effectively reduces the entrance of water. When mortar joints are used, nonstaining cement offers the best protection against staining.

GRANITE

Physical Characteristics

Granite is a crystalline igneous rock composed mainly of feldspar and quartz with lesser amounts of mica and hornblende. Among building stones, granite is one of the hardest and most difficult to cut and finish. However, for exterior applications it is one of the most enduring. See Figure 4-23 for granite installation.

FIGURE 4-23 Wyndham Hotel, Dallas, TX: Granite—Sunset Red, polished finish. Dahl/Braden & Chapman, Architects. (Courtesy of Cold Spring Granite Company.) This illustration also appears in the color section.

Colors

Depending on the quarry, domestic granite can be obtained in the following colors: white, gray, buff, beige, pink, red, blue, green, brown, and black. Granite in varying colors is likewise available from Canada and abroad.

When a block of granite is taken from a quarry it may be cut parallel to one of three axes. Depending upon which axis it is cut, slight variations in the appearance will result.

Classification

The National Building Granite Quarries Association classifies architectural granite as follows:

Building Granite: Granite used either structurally or as a veneer for exterior or interior wall facings, steps, paving, copings, or other building features.

Masonry Granite: Granite used in large blocks for retaining walls, bridge piers, abutments, arch stones, and similar purposes.

Finishes

Finishes that are most commonly available are as follows:

Polished: Mirror gloss, with sharp reflections.

Honed: Dull sheen, without reflections.

Fine Rubbed: Smooth and free from scratches; no sheen.

Rubbed: Plane surface with occasional slight "trails" or scratches.

Shot Ground: Plane surface with pronounced circular markings or trails having no regular pattern.

Thermal: Plane surface with flame finish applied by mechanically controlled means to ensure uniformity. Surface coarseness varies, depending upon grain structure of granite.

Sand-Blasted Coarse Stippled: Coarse, plane surface produced by blasting with an abrasive; coarseness varies with type of preparatory finish and grain structure of granite.

Sand-Blasted, Fine Stippled: Plane surface, slightly pebbled, with occasional slight trails or scratches.

8-Cut: Fine bushhammered; interrupted parallel markings not over 3/32 inch apart; a corrugated finish smoother near arris lines and on small surfaces.

6-Cut: Medium bushhammered finish, similar to but coarser than 8-cut, with markings not more than 1/8 inch apart.

4-Cut: Coarse bushhammered finish with same characteristics as 6-cut, but with markings not more than 7/32 inch apart.

Sawn: Relatively plane surface with texture ranging from wire sawn, a close approximation of rubbed finish, to shot sawn, with scorings 3/32 inch in depth. Gang saws produce parallel scorings; rotary or circular saws make circular scorings. Shot-sawn surfaces are sandblasted to remove rust stains and iron particles.

Special finishes of many kinds are also available from some producers to meet special design requirements. Samples should be obtained to ascertain how the granite species will appear with the special finish.

Minimum Thickness

The thickness of exterior veneer panels have been discussed under Veneer Stone. For bushhammered or pointed finish, a minimum thickness of 4 inches is required.

Joint Widths

Granite joints are usually 1/4 inch thick. Smaller joint widths where desirable for design purposes may be obtained, but the costs will increase to achieve this end result.

Setting

In addition to the proprietary systems noted under "Veneer Stone" under GENERAL, granite veneer can be set and anchored with the anchors described under "Anchors and Anchorage," and Figure 4-21.

When granite veneer is set, the vertical loads are typically carried on discontinuous angle supports arranged so that the angle is centered at the vertical joint between adjacent stones. Pads or buttons of plastic, neoprene or lead, generally 1 x 1 inch by the joint thickness, are used at each horizontal joint to maintain the joint width until the sealant or mortar sets.

Checklist

1. The quarry must have sufficient stone of the color selected to satisfy the needs of the project.
2. The stone selected must be capable of withstanding the rigors of the weather at the project site.
3. Veneer thickness is a function of panel size especially with regard to handling, transportation, erection, and structural adequacy.
4. Detail the installation to inhibit moisture movement. Use waterproofing on back-up materials to prevent staining. At bases, detail the waterproofing as shown in Figure 4-22.
5. Mortar when used should be a nonstaining type.
6. Obtain sample to verify color, finish and texture variations.
7. Specify shop and erection drawings together with structural calculations where necessary for anchorage system.
8. When storing stone at project site prior to setting, make certain that temporary wood strips or other parting devices do not stain or disfigure the stone.
9. During construction, unfinished stonework should be protected from rain or snow by plastic or nonstaining coverings.

LIMESTONE

Physical Characteristics

Limestone is a sedimentary rock composed chiefly of calcium carbonate. The chief varieties are oolitic limestone and dolomitic limestone. The latter has a high magnesium carbonate content.

Oolitic limestone, which is quarried mainly in Indiana, is a finely divided calcium carbonate formed of shells and shell fragments that were broken, crushed and ground, and then redeposited in a shallow sea. Oolitic limestone is characteristically a freestone (a stone that may be cut freely in any direction without fracture or splitting) without pronounced cleavage planes, possessing a remarkable uniformity of composition, texture, and structure. This attribute lends itself to ease in machining and provides flexibility of shape and texture at low cost.

Dolomitic limestone is richer in magnesium carbonate (about 10 to 15% MgO) and frequently somewhat crystalline in character. It has higher compressive and modulus of rupture strengths than oolitic limestone. The dolomitic varieties are mined mainly in Minnesota and contain a broader range of color and texture than oolitic limestone. Some of these stones are classified as marble by the Marble Institute of America.

Colors

Oolitic limestone is available in two colors as classified by the Indiana Limestone Institute—buff and gray. Buff varies from a light, creamy shade to a brownish buff. Gray varies from a light, silvery gray to shades of bluish gray.

Dolomitic limestone is available in gray tones, pinks, light to dark brown, and bluish-gray.

Classification and Uses (Oolitic Limestone)

The Indiana Limestone Institute classifies oolitic limestone by the degree of fineness of the grain particles and other natural characteristics that make up the stone. Structurally, the grades are essentially identical.

Oolitic limestone contains a few distinguishable calcite streaks or spots, fossils of shelly formations, pit holes, reedy formations, open texture streaks, honeycomb formations, iron spots, travertine-like formations, and grain-formation changes. Based on these characteristics, oolitic limestone is graded by the Indiana Limestone Institute as follows:

1. *Select:* Fine to average-grained stone having a controlled minimum of the above characteristics.
2. *Standard:* Fine to moderately large-grained stone permitting an average amount of the above characteristics.

3. *Rustic:* Fine to very coarse-grained stone permitting an above-average amount of the above characteristics.

4. *Variegated:* An unselected mixture of Grades 1 through 3, permitting both the buff and gray colors.

Select Grade is generally used and confined to those areas of a building where: (1) cost is not a prime factor; (2) smooth machine finishes are desired; and (3) molded, carved and sculptured detail is intended. As such it is well suited for lower-story work, entrances, and similar details where it is readily observed.

Standard Grade is most commonly used where a general uniformity of texture is desired and cost is a consideration. It is particularly useful for commercial and monumental building for the ashlar stone work where Select Grade is neither appropriate nor warranted and where the slight variation in texture does not detract from the overall design intent. Standard Grade is also adaptable to the smooth finishes.

Rustic Grade is adapted to the less formal types of architectural design where a wide range of color tone (either buff or gray) and texture are appropriate.

Variegated Grade is used where mixtures of color tones (buff and gray) and texture are appropriate, especially for the rougher finishes such as "chat-sawed," "shot-sawed," and "sand-sawed."

Finishes (Oolitic Limestone)

Smooth: Produced by a planer or grinder, or by a circular sander using a carborundum-faced sanding disk. Produces a relatively smooth surface with a certain amount of texture.

Plucked: A machine finish obtained by rough planing the surface, thus breaking or plucking out small particles and resulting in an interesting rough texture.

Machine-Tooled: Parallel concave grooves are cut into the stone, in 4, 6 or 8 bats (grooves) to the inch. Depth of groove varies between $1/32$ to $1/16$ inch.

Chat-Sawed: This finish is produced during the gang sawing operation using a coarse pebbled surface resembling sand blasting.

Shot-Sawed: By using steel shot in the gang-sawing operation, a coarse, uneven finish is produced, ranging from a pebbled surface to one ripped with irregular, roughly parallel grooves. The steel shot rusts during the process, allowing a certain amount of rust stains to develop and adding permanent brown tones.

Split Face: A rough, uneven, concave convex finish is produced through a splitting action.

Rock Face: Similar to split face except that the face of the stone is dressed by machine or hand to produce a bold, convex projection along the face of the stone.

Finishes (Dolomitic Limestone)

Split Face: Same as oolitic limestone.

Gang Sawed: Consists of nearly parallel striations up to about $1/16$ inch deep.

Shot-Sawed: Same as oolitic limestone.

Carborundum: Same as oolitic limestone.

Planer: Similar to Smooth Finish of oolitic limestone.

Sand-Rubbed: Fine sand used as abrasive to produce a smooth finish.

Honed: Smoothest flat plane finish without a surface shine.

Polished: Shiny and reflective surface.

Tapestry: A heavy, sand-blasted finish that emphasizes the petrified plant life of the stone.

Split Rusticated Finish: Produced by a combination of sawing and splitting.

Panel Sizes and Thicknesses

Dolomitic limestone has greater crushing and tensile strengths than oolitic limestone. With oolitic limestone, the Indiana Limestone Institute (ILI) recommends certain panel sizes and thicknesses for efficient fabrication and handling, as shown in Table 4-17.

TABLE 4-17
ILI MAXIMUM PANEL SIZES

Height (Feet)	Width (Feet)	Thickness (Inches)
5	3	2
9	4	3
11	5	4
14	5	5
18	5	6

Thin veneers under 2 inches in thickness are not usual with oolitic limestone. It is best to review the design requirements of a project with a producer to properly engineer the panel dimensions, after considering finishes, transportation, handling, and fabrication techniques.

Checklist

See Checklist under Granite.

MARBLE

Physical Characteristics

Marble is a metamorphic rock, generally a metamorphosed limestone composed of crystalline and compact varieties of calcium carbonate and/or magnesium carbonate. Its crystalline structure permits polished marble to gleam, since light penetrates a short distance and is then reflected from the crystalline structure. See Figures 4-24, 4-25, and 4-26 for marble installations.

Marble varies greatly in its ability to withstand at-

FIGURE 4-24 National Gallery of Art, East Building, Washington, DC: Tennessee marble. I.M. Pei & Partners, Architects. This illustration also appears in the color section.

mospheric durability. Acid rain, industrial pollution, and automobile exhaust fumes may cause severe deterioration to the point where the material will be etched, flake, and ultimately crumble and disintegrate.

Colors

The intrusion of other substances and impurities during the formation of marble results in a very colorful and

FIGURE 4-25 Beinecke Rare Book & Manuscript Library, Yale University: white marble. Skidmore, Owings & Merrill, Architects. (Photographer, Ezra Stoller © ESTO.)

FIGURE 4-26 Interior view of Figure 4-25 showing translucent characteristics of marble. (Photographer, Ezra Stoller © ESTO.) This illustration also appears in the color section.

veined stone that makes marble one of the more desirable decorative building stones, that can take a high polish to reflect these characteristics.

Marble can range from white to black. Domestic marble colors include tan, pink, rose, red, brown, green, gray, and blue-gray. Marble is also available from foreign countries in a multitude of colors.

Classification

Marble is classified by groups by the Marble Institute of America (MIA) as follows:

Group A: Sound marbles and stones with uniform and favorable working qualities.

Group B: Marble and stones similar in character to Group A, but working qualities somewhat less favorable; occasional natural faults; limited amount of waxing* and sticking[†] necessary.

Group C: Marbles and stones of uncertain variation in working qualities; geological flaws, voids, veins, and lines of separation common; which require repair by waxing,* sticking,[†] and filling; liners[‡] and other forms of reinforcement freely employed when necessary, before or during installation.

*Waxing: Finishing by filling the natural voids in marble with color-blended material.
[†]Sticking: Process of cementing together broken slabs or pieces of marble.
[‡]Liners: Structurally sound sections of marble that are cemented to the back of marble veneer to give strength or additional bearing surface.

Group D: Marbles and stones similar to Group C but with a larger proportion of natural faults and maximum variations in working qualities. This group comprises many of the highly colored marbles prized for their decorative qualities. The same methods of repair as used for Group C are required.

The classification of marble into these four groupings is based on experience gained over the years by marble suppliers and installers. The basis of this classification is the characteristics encountered solely in finishing and indicates what method of finishing is considered proper and acceptable.

While the C and D groups of marbles are more decorative by virtue of their veining and geological flaws, they are not necessarily suited for all building applications, especially exteriors. The Group A marbles are the soundest of all and are the ones most usually used for exterior applications.

Finishes

Marble can be finished in a variety of ways to enhance its appearance; this can range from polished to rough, as noted in Table 4-18

TABLE 4-18
MARBLE FINISHES

Polished	A mirror-like, glossy surface that brings out the full color and character of the marble. Produced by polishing a honed surface with a textile buffer and fine abrasives.
Honed	A velvety-smooth surface with little or no gloss produced by machine or hand rubbing with special abrasives.
Grit	A smooth, dull finish obtained by rubbing with an abrasive grit.
Sand, blown	A smooth, matte surface achieved by a light sand blasting.
Sand, wet	A smooth surface obtained by rubbing with a machine using sand and water as abrasives.
Sanded	A smooth, dull finish produced by rubbing with sand.
Natural	A moderately rough, textured face produced by sawing with sand and water as abrasives.
Split Face	A rough natural face of stone produced by machine splitting.

Applications

Since marble is one of the most colorful of the building stones, it is widely used in all types of building applications. However, the veining and geological faults which create the beauty of marble inhibit its exterior use. Repair by way of sticking waxing and reinforcement allow Groups C and D marble to be used for interiors but rarely for exterior applications.

Polished finishes are restricted to a handful of marbles when used for exteriors. A honed finish is recommended for flooring.

In addition to the use of marble for exterior and interior walls, it can be used for copings, sills, saddles, treads and risers, paving, flooring, toilet partitions, and countertops.

Split-face finished marble is used for masonry ashlar work where the cut stone may have a bed about 4 inches deep, a height of from 2 to 8 inches, and lengths of about 12 to 48 inches.

Thin Veneer Panels

Marble exterior facings can vary in thickness from $7/8$ to 2 inches. The required thickness will depend on the panel size, anchoring provisions, the handling requirements, and the local building code. In general, the greater thicknesses are desirable for high-rise structures, or where the panel faces are large, or where the weather is severe. Thinner panels are adequate where the panel sizes are small or for low storefront work. However, it is prudent to check the thickness to be detailed with several marble suppliers and installers.

Checklist

1. See checklist for Granite.
2. For thin veneers, provide a space between marble and back-up for air circulation.
3. Where solid grouting is required between marble and back-up, use nonstaining, waterproofed mortar utilizing portland cement.

SANDSTONE

Physical Characteristics

Sandstone is a sedimentary rock usually consisting of quartz cemented with silica, iron oxide, or calcium carbonate. Sandstone is durable and has a very high crushing and tensile strength. Bluestone and brownstone are sandstones, named for their colors.

Colors

Sandstones come in a variety of colors. Sandstones of pure silicon dioxide are white. With the iron oxide impurities, sandstone colors are available in grays, yellows, browns, reds, tans, and pinks.

Finishes

Sandstone is available in split-face, rustic-face, pitched-face, chat-sawed, sand-sawed, and smooth-sawed finishes.

Sizes and Uses

A good deal of sandstone is used as rubble stone masonry and as split-face ashlar stone masonry in sizes varying in bed width from 3- to 5-inches, in height from 2 to 12 inches, and in length from 8 to 36 inches.

TABLE 4-19
MAXIMUM FACE SIZE AND
THICKNESS OF SANDSTONE

Face Size	Thickness
2 x 4 feet	2¼ inches
6 x 3½ feet	3 inches
8 x 3½ feet	3½ inches
8 x 3½ feet	4 inches

Veneer panels are available from some suppliers in the sizes shown in Table 4-19.

Rubble stone masonry and split face masonry are set in mortar and are usually wall bearing. Veneer stone panels are generally installed with conventional stone anchors and mortar joints.

SLATE

Physical Characteristics

Slate is a very fine-grained metamorphic rock derived from sedimentary rock shale. It is characterized by an excellent parallel cleavage entirely independent of the original bedding, by which cleavage the rock may be split easily into relatively thin slabs.

Color

Slate owes its colors to ingredients other than shale. Carbonaceous or iron sulfides produce the dark colors of black, blue, and gray. Iron oxide accounts for the reds and purples; and chlorite the green.

Select slate is uniform in color. Ribbon slate contains bands or ribbons of darker color and sometimes ribbons of different colors. The colors may be unfading or weathering (color changes with exposure of weather).

Finishes

Slate is available in several finishes as follows:

Honed: A semipolished finish, smoother than sand-rubbed but without a high sheen surface.

Sand Rubbed: Surface is rubbed to remove all natural clefts using the equivalent of a 60-grit abrasive.

Natural Cleft: This is the natural split or cleaved face. It is moderately rough with some textural variation.

Semi-Rubbed: Approximately 50% of natural cleft face removed.

Gauging: A grinding process used on backs of natural cleft surfaces to produce a more even thickness for better fit.

Uses

Slate is used for a variety of building areas both interior and more recently as exterior wall and column facings. Some areas of use are as follows:

Interior walls
Flooring
Paving
Flagging
Treads and risers
Stools
Shelving
Blackboards
Electrical panel boards
Fireplaces
Bases
Roofing shingles

Sizes and Thicknesses

Slate is not used for ashlar or rubble masonry. Black slate is used primarily as a facing veneer on exterior and interior walls. All varieties are used as flooring (see Chapter 9) and paving (see Chapter 2). For exterior face veneers sizes of up to 5 feet x 8½ feet x 1½ inches have been used successfully. Smaller exterior panels have utilized 1-inch thick slate. Limitations for individual panels are not over 9½ feet in length and not over 5 feet in width.

Setting

Exterior Wall Panels: Each slate panel should be anchored separately by nonferrous ¼-inch diameter wire turned down 1 inch into a ⅜ inch round, 1½ inch deep hole in the edge. A minimum of four anchors is recommended for panels up to 12 square feet, with two additional anchors for every additional 6 square feet of panel. Doweling natural cleft slate to slate is not recommended. Relief angles and occasionally liners are recommended at normal floor-line distances. The back of the slab should be a minimum of 1 inch from back-up, with an air cavity, and with portland cement spots 6 x 6 inches centered 18 inches apart.

Exterior Jointing: For natural cleft finishes, exterior joints should be ⅜ to ½ inch wide because of the variation in the texture and lippage of adjacent panels.

Interior Wall Panels: Setting is similar to exterior panels except that in some cases, small panels ¼ inch thick may be set in mastic and with butt joints.

TRAVERTINE

Physical Characteristics

Travertine is a sedimentary rock and a variety of limestone, which is deposited from solution. It is a product of chemical precipitation from hot springs. Travertine is cellular with the cells usually concentrated in thin layers that display a stalactic structure. Some that take a polish are sold as marble and may be classified as "travertine marble" under the class of commercial marble. As a result of its manner of formation, travertine is characterized by many irregular cavities. Figure 4-27 shows the travertine facade of the LBJ Library, Austin, Texas.

Colors

Travertine is available in grays, pinks, rose, and shades of tan. Some travertine has large fissures, and others less noticeable fissures.

Travertine is produced in the United States, but the more exotic colors and formations are quarried near Rome in Italy. However, for a major project it would

FIGURE 4-27 LBJ Library, Austin, TX: Travertine facade. Skidmore, Owings & Merrill, Architects. (Photographer, Ezra Stoller © ESTO.) This illustration also appears in the color section.

be essential to visit the quarries or obtain 2½- x 5-foot samples to select the specific material desired. One need only examine the many structures at Lincoln Center in New York to visualize the differences between various travertines from different quarries.

Finishes, Uses, Setting

Travertine can take a high polish and is finished, used, and set like marble. When travertine is used for structures such as countertops, the fissures may be filled with chips of travertine and colored mortar to match the adjacent areas.

Checklist

1. See checklist under Granite.
2. For large projects, an inspection of the quarry to make selections and ensure an adequate supply is recommended.
3. Verification of specified material by submission of large panel samples (2½ x 5 feet) is suggested.

5

METALS

FABRICATION PROCESSES

Good architectural design and detailing require not only a knowledge of the characteristics of the metals used in ornamental metal work but an understanding of how they are worked, shaped, and formed in the fabrication process. Each metal by virtue of its physical characteristics has its own peculiarities, and the ease or difficulty in forming is reflected in its cost-in-place. Obviously, by understanding fabrication processes, economies may be introduced by proper selection of standard sizes and shapes and by detailing.

The following definitions of shaping and forming terms will aid in understanding the fabrication process:

Blanking: In sheet metal work, the cutting out of a piece of metal, usually by means of a press.

Braking: A mechanical bending operation usually performed on sheets and plates.

Broaching: Finishing the inside of a hole to a shape other than round.

Casting: An article formed by solidification of molten metal in a mold.

Cold Drawing: Drawing metal through a die without the application of heat.

Cold-Rolled: Metal rolled at room temperature, below the softening point, usually harder, smoother, and more accurately dimensioned than hot-rolled material.

Cold Working: The process of changing the form or cross-section of a piece of metal at a temperature below the softening point.

Countersinking: Beveling the edge of a hole for the reception of the head of a bolt, rivet, or screw.

Drawing: Forcing metal to flow into a desired shape without melting, by pulling it through dies.

Embossing: Development of a raised design on a metal surface by die pressure or by stamping or hammering on the reverse surfaces.

Extrusion: Forcing a molten metal through a die by pressure.

Fluting: Formation of a semicircular groove or series of grooves in sheet metal.

Forging: Heating and hammering or pressing metal into a desired shape.

Forming: Pressing metal into shape by mechanical operations other than machining.

Hot-Forming: Working operations such as bending, drawing, forging, piercing, and pressing performed at temperatures above the recrystallization temperature of the metal.

Lost-Wax A process using patterns of wax which are
Process: melted and drained from the mold before the metal is poured; used in the making of castings involving undercuts and other complications.

Milling: Removal of metal to develop a desired contour by means of a revolving cutting tool.

Perforating: Punching or drilling multiple holes in sheet metal, blanks, or formed parts.

Pressing: Forcing metal to conform to the shape of a die by means of pressure.

Punching: Forcing a punch through metal into a die, forming a hole the shape of the punch.

Reaming: Enlarging a round hole in a piece of metal by means of a revolving edged tool.

Riveting: Forming a connection between two or more pieces of metal by passing a rivet through aligned holes and upsetting to form a head.

Rolling: Shaping metal, either hot or cold, by passing it between revolving rolls set to a predetermined distance apart. Rolls may be flat, or shaped as desired.

Seaming: Uniting the edges of a sheet or sheets by bending over or doubling, and pinching the edges together.

Shearing: Cutting metal by the action of two opposing passing edges.

Spinning: Shaping sheet metal by bending or buckling it under pressure applied by a smooth hand tool or roller while the metal is being revolved rapidly.

Stamping: Bending, shaping, cutting out, indenting, embossing, or forming metal, either hot or cold, by means of shaped dies in a press or power hammer.

Straightening: Eliminating deformations by pressing, rolling, or stretching.

Stretcher A process of stretching a metal sheet to
Leveling: produce a straight, flat surface.

Swaging: Surface working of a forging (either hot or cold) by means of repeated blows, usually between dies.

Tapping: Cutting internal threads in a punched or drilled hole.

Tempering: A heat treatment whereby metal is brought to a desired degree of hardness and elasticity.

Tumbling: A process of cleaning metal articles by placing them in a revolving container, with or without cleaning material.

Turning: Removal of metal by means of an edged cutting tool while the piece is being revolved about its axis.

Upsetting: Building up or thickening the section of a piece of hot or cold metal by shortening the piece by axial compression.

Work Hardening: The increase in hardness developed in a metal as a result of cold working.

STANDARD STOCK MILL PRODUCTS

For most architectural applications there are standard stock metal items, produced in the form of sheet, plate, strips, bars, rods, wire, or extrusions, that are used to develop either simple shapes and forms, or combinations of these mill products that are used to develop complex shapes and forms. The following definitions are representative of standard stock items of most metals.

Angle: A section of metal rolled, drawn or extruded through L-shaped rolls or dies.

Bar: Round, square, rectangular, hexagonal, or other solid stock of drawn, rolled, or extruded metal.

Channel: A rolled, drawn, or extruded metal section having a U shape.

Flat: A rectangular bar whose width is greater than its thickness.

Pipe, Round: A hollow, round section of metal, the size of which is determined by the nominal inside diameter in inches.

Pipe, Square: A hollow, square section of metal, the size of which is determined by the nominal outside diameter in inches.

Plate: A flat piece of metal. Various metals are defined as plate by the following thickness criteria:

Aluminum	¼ inch or more
Copper	0.188 inch or more
Steel	³⁄₁₆ inch or more
Stainless steel	³⁄₁₆ inch or more

Rod: See Bar.

Round: A cylindrical rod (see Rod).

Sheet: A thin, flat piece of metal, the thickness of which is thinner than plate (see Plate).

Strip: Narrow metal sheets either produced in coil form or cut to length and finished flat.

Tubing: A hollow section of metal which may be round, square, rectangular, hexagonal, octagonal, or other shape, measured by the external size in inches and wall thickness by gage or decimals of inches.

Wire: A small-diameter rod measured by gage, produced by a drawing process (see Rod).

JOINING

Both ornamental and utilitarian metal work are fabricated in units and assembled or joined in the shop into sections. The sections joined together in the field into the finished work. Joining is an important aspect of metalwork and may be performed in a number of different ways, depending upon the type of metal and the specific problem. Typical joining procedures include bolting and riveting, welding, brazing, and soldering.

Bolting and Riveting

Bolting and riveting are similar operations where two or more pieces of metal are connected by a bolt or rivet that passes through holes. Bolting may be temporary or permanent. Riveting is a permanent connection, where one end of the shank is upset by hammer blows to form a head. During this process the shank expands to fill the holes fully, thus forming a strong, tight joint. Riveting is rarely used now for structural steel.

Welding

Welding is a process of joining metals by applying heat and pressure, with or without filler material, to produce an actual union through fusion. There are a number of welding processes employed as follows:

Carbon Arc: An electric arc process wherein a carbon electrode is used and fusion is produced by heating. Pressure may or may not be applied, and filler metal may or may not be used.

Electric Arc: A process wherein a metal electrode is used which supplies the filler material in the weld and the heat to produce the fusion.

Forge Welding: A process wherein fusion is produced by heating in a forge or furnace and applying pressure or blows to the work.

Fusion Welding: A process of welding without pressure, in which a portion of the base metal is melted. It is usually accomplished by gas flame or electric arc heating.

Gas Welding: A process of welding wherein fusion is produced by heating with gas (acetylene, hydrogen), with or without pressure and with or without filler material.

Resistance Welding: A process of welding accomplished by placing the work to be joined under pressure in a machine, then applying an electric charge through the joint, the resistance of which produces heat to fuse the joint.

Welding sometimes imparts distortion, brittleness, or changes in strength and ductility at the joint. To overcome these deficiencies, cold working and annealing are often necessary to restore the original working characteristics. With nonferrous metals such as aluminum the factor of color may have an important bearing upon the choice of the proper welding process and the proper electrode or filler metal.

Brazing

Brazing is a process wherein a molten filler metal is used to join metal parts. The filler metal has a melting point below that of the metals to be joined. Brazing is accomplished at temperatures above that of soldering (800°F) and below that of welding. Heating for brazing may be accomplished by dipping the parts into a bath of the molten alloy, by heating with torches, furnace heating, or by electrical resistance.

Brazing materials must be selected to offer corrosion resistance between base metals and filler, or, where a color match is required, between the base metal and the brazing metal, or where the strength of the joint will not be impaired by the brazing metal.

Brazing materials consist of the following:

Aluminum-silicon: For brazing aluminum.

Copper-phosphorous: For brazing copper and copper alloys.

Silver: For brazing ferrous metals, copper, and copper alloys.

Copper and copper zinc: For brazing ferrous metals, copper, and copper alloys.

Heat-resisting alloys: For brazing ferrous metals.

Soldering

Soldering is a process similar to brazing with the filler metal having a melting temperature range below 800°F. Since soldering temperatures are low, there is no alloying of the base metal and the solder. As a result the base metals are usually stronger than the joint. Where the strength of the joint in sheet metal work is to be improved, it is advisable to reinforce the seams by crimping, interlocking, riveting, or bolting before soldering, depending upon the solder to make the joint tight. Aluminum is not easily soldered. See Aluminum for further information.

HEAT TREATING

Heat treating is employed in metal working to induce certain properties or to relieve stresses and strains after certain metal-working processes have been performed. Heat-treating processes include the following:

Annealing: A heating and cooling operation performed on metal in the solid state involving cooling at a relatively slow rate. The process generally results in reducing hardness, improving machinability, facilitating cold working, removing stresses, and altering the ductility and toughness of metals. The temperature of the operation and the rate of cooling will depend upon the specific metal and the purpose of the annealing process.

Tempering: A specific heat treatment whereby metal is brought to a desired degree of hardness or softness.

SURFACE WORKING

Surface-working processes involve the application of certain operations to the surface of metals to alter their appearance and include the following:

Blasting, Cleaning: Obtaining a mottled or pebbled surface by means of blasting the surface with sand, grit, or steel shot through a nozzle by air pressure.

Brushing, Buffing, Polishing: Producing smooth,

satin, bright, or buffed finishes by means of wheels on high-speed lathes, by belts on sanding wheels, or by disks and wheels on hand buffers and grinders.

Etching: A process of chemical etching the surface by means of acid or alkali solutions to obtain decorative architectural effects.

Grinding: A surface texture is obtained by varying the grit in a grinding wheel or disk.

Hammering: Metal surfaces can be altered by indentation through hammering or peening to obtain the desired degree of surface alteration.

ALUMINUM

Aluminum is a nonmagnetic silvery white metal that is easily formed in the fabrication process by extrusion, rolling, drawing, stamping, casting, and forging. It is light in weight (about one-third the weight of iron) and has a melting point of about 1200°F. The thermal coefficient of expansion is 0.0000128, and the tensile strength is 22,000 psi.

Since aluminum unalloyed is deficient in strength and soft, it is rarely used in building construction in its refined state. By alloying it with other elements, its physical characteristics are enhanced considerably.

Aluminum alloys are available in the form of bar, castings, extrusions, forgings, pipe, plate, sheets, structural shapes, tubes, and wire. Aluminum provides a high degree of corrosion resistance; since the products of corrosion are white, staining of adjacent surfaces is reduced.

ALUMINUM ALLOYS

Alloy Classification System

The Aluminum Association classifies aluminum alloys on the basis of the major alloying metal. The most widely used wrought aluminum alloys in building construction are shown in Table 5-1.

TABLE 5-1

ALUMINUM ALLOYS

Alloy	Major Alloying Metal
Series 2000	Copper
Series 3000	Manganese
Series 4000	Silicon
Series 5000	Magnesium
Series 6000	Magnesium and silicon
Series 7000	Zinc

ALLOY SELECTION GUIDE

The following is a guide to the selection of aluminum alloys.

Sheet and Plate

3003	Most widely used for roofing sheet metal applications; where anodizing is not required. Assumes a slight yellowish cast when anodized.
Alclad[a] 3004	A clad aluminum sheet used for standard corrugated, ribbed, or V-beam section and various embossed patterns for industrial roofing and curtain wall sheets.
5005	Commonly used for low-cost, all-purpose sheets with good formability and finishing characteristics. Has good anodized appearance match with alloy 6063 extrusions.
5050	An alloy with good corrosion resistance and stronger than alloys 3003 and 5005. Produces a comparatively clear, white appearance after anodizing.
5052	Good workability and excellent corrosion resistance. Assumes yellowish cast when anodized. Used for venetian blinds and weatherstripping.

[a] The term *alclad* refers to a protective cladding of aluminum that is applied to thin sheets of a core alloy to improve its corrosion resistance. When buffing clad products, exercise care not to wear through cladding. Do not grind. The cladding may be of the same or different alloy than the core.

6061 — Most economical and versatile of the heat-treatable alloys. Used for roll-formed structural shapes and applications requiring high strength. Good anodized appearance match with alloy 6061 extrusions, but may take on yellowish cast when anodized.

Extrusions

6061 — Used principally for extruded structural shapes, pipe and tubing, or applications requiring high strength. For color anodizing matching, use in conjunction with alloy sheet 6061.

6063 — Most commonly used extrusion alloy for building and ornamental metal products. For anodized color matching, use alloy sheet 5005.

Castings

43 — Used for hardware and ornamental metalwork where strength is not critical; turns gray after anodizing.

214 — Stronger than alloy 43. Best appearance match with 6063 extrusions and 5005 sheets, when anodized.

356 — High-strength casting alloy; turns gray after anodizing.

Special Alloys

[b] For proprietary alloys (both sheet and extrusions) designed for quality applications where excellent anodized appearance is critical, it is best to consult with the major aluminum producers.

Porcelain Enamel Alloys

1100 Sheet — A commercial pure aluminum, with low mechanical properties and good workability. Specify PE grade.

3003 Sheet — Higher mechanical properties than 1100 alloy. Has excellent framing characteristics. Use PE Grade 3003.

6061 Sheet & extrusion — Stronger and harder than alloys 1100 and 3003, combines good mechanical properties with good resistance to corrosion.

7104 Extrusion — Allows maximum freedom of design for extruded shapes. It combines the extrusion characteristics of alloy 6063 with higher mechanical properties and good workability.

Casting Alloy43 — A general-purpose casting alloy.

Casting Alloy356 — Good casting characteristics and heat treatable to develop maximum mechanical properties.

Specialty Porcelain Enamel Alloys

Improved alloys are constantly being discovered as a result of research, commercial practice, and metallurgical advances. Consult porcelain enamel fabricators for most recent development of porcelain enamel aluminum alloys.

TEMPER AND HEAT TREATMENT

The tempering of aluminum alloys is achieved by mechanical or thermal treatment or by a combination of both, to improve strength or hardness. By alloying aluminum with other elements, strength and hardness are imparted to the pure aluminum. Further strengthening is achieved by classifying the alloys into two categories: non-heat-treatable and heat-treatable.

Non-heat-treatable alloys, Series 1000, 3000, 4000, and 5000, are further strengthened by cold working, denoted by the "H" series of tempers.

Heat-treatable alloys, Series 2000, 6000, and 7000, which contain copper, magnesium, zinc, and silicon either singly or in various combinations, will respond to pronounced strengthening when heat treated.

Heat treatment or solution heat treatment is an elevated temperature process designed to put the soluble alloying elements in solid solution. This is followed by rapid quenching, usually in water, which momentarily "freezes" the structure and for a short time renders the alloy very workable.

Artificial aging or precipitation hardening is achieved by heating for a controlled time at slightly elevated temperatures.

TABLE 5-2.
BASIC TEMPER DESIGNATIONS

F	As fabricated—Applies to products of shaping processes in which no special control over thermal conditions or strain hardening is employed.
O	Annealed—Applies to wrought products which are annealed to obtain the lowest-strength temper, and to cast products which are annealed to improve ductility. The O may be followed by a digit other than zero.
H	Strain-hardened (wrought products only)—Applies to products which have their strength increased by strain hardening, with or without supplementary thermal treatment to produce some reduction in strength; always followed by two or more digits.
W	Solution-heat-treated—Unstable temper, applies to alloys which spontaneously age at room temperature after solution heat treatment.
T	Thermally treated to produce stable tempers other than F, O, or H; followed by one or more digits.

TABLE 5-4
SUBDIVISION OF T TEMPERS

T1	Cooled from an elevated temperature-shaping process and naturally aged to a substantially stable condition.
T2	Cooled from an elevated temperature-shaping process, cold worked, and naturally aged to a substantially stable condition.
T3	Solution heat-treated, cold worked, and naturally aged to a substantially stable condition.
T4	Solution heat-treated and naturally aged to a substantially stable condition.
T5	Cooled from an elevated temperature-shaping process and artificially aged.
T6	Solution heat-treated, then artificially aged.
T7	Solution heat-treated and then stabilized.
T8	Solution heat-treated, cold worked, then artificially aged.
T9	Solution heat-treated, artificially aged, then cold worked.
T10	Cooled from an elevated temperature-shaping process, cold worked, then artificially aged.

Basic Temper Designations

A temper designation system, based on heat treatment, aging, and annealing, has been formulated consisting of letters and number designations. Table 5-2 shows the basic temper designations which are usually added as a suffix to the alloy type, for example, 6063-T5. Table 5-3 shows the subdivision of H tempers and Table 5-4 shows the subdivision of T tempers.

For fabricated products, manufacturers usually purchase aluminum stock shapes that are tempered to the requirements necessary for the specific products. When designing building components, consult with the aluminum producer and fabricator to ascertain the required alloy tempers for the specific needs.

TABLE 5-3
SUBDIVISION OF H TEMPERS

H1	Strain hardened only. The digit indicates degree of strain hardening.
H2	Strain hardened and partially annealed.
H3	Strain hardened and stabilized.

CLEAR AND COLOR-ANODIZED FINISHES

Aluminum is unique among the architectural metals with respect to the wide variety of finishes including the integral clear and color-anodized finishes available as a result of research and development.

There are two important advantages of finished aluminum to the architect: (1) appearance, and (2) protection. These are derived on the basis of certain pretreatments which include: (1) mechanical finishes (ranging from polished to satin, bright, matte, and pebble-textured), and (2) chemical finishes (ranging from caustic etch to acid etch, brightened and chemical conversion coatings). Following these pretreatments (except for chemical conversion coatings), the aluminum may then be anodized to obtain a clear coating or an integral color-anodized coating.

Aluminum may be used without an anodized finish since exposure to the air will result in the formation of a thin, protective oxide film. Prolonged exposure to the weather, however, will result in roughening of the surface with attendant chalkiness.

Anodizing, which is a controlled oxidizing process, produces a thicker, denser and harder oxide coating that adds to the durability of the metal. It is recommended that for exterior architectural applications, only aluminum which has been anodized be used.

Mechanical Finishes

Mechanical pretreatment is a process whereby the surface characteristics of the metal are induced by mechanical means. These characteristics may be achieved by the original mill production processes such as rolling and extruding or by subsequent polishing, grinding, brushing, or blasting techniques. The mechanical processes used and their designations have been classified by the Aluminum Association and are shown in Table 5-5.

The mechanical finishes are performed on individ-

TABLE 5-5
MECHANICAL FINISHES (M)
ON ALUMINUM

As Fabricated
M10	Unspecified
M11	Specular as fabricated
M12	Nonspecular as fabricated
M1X	Other (to be specified)

Buffed
M20	Unspecified
M21	Smooth specular
M22	Specular
M2X	Other (to be specified)

Directional Textured
M30	Unspecified
M31	Fine satin
M32	Medium satin
M33	Coarse satin
M34	Hand rubbed
M35	Brushed
M3X	Other (to be specified)

Nondirectional Textured
M40	Unspecified
M41	Extra-fine matte
M42	Fine matte
M43	Medium matte
M44	Coarse matte
M45	Fine shot blast
M46	Medium shot blast
M47	Coarse shot blast
M4X	Other (to be specified)

ual parts or components, generally a hand operation, although mechanical equipment is involved. These operations may include buffing wheels, belt polishers, sanders, and blasting equipment.

Prior to specifying any mechanical finish, it is best to consult the aluminum producer and product fabricator to ascertain whether certain items are capable of receiving specific mechanical finishes. For example, small-radius inside corners and inaccessible areas on extruded or formed shapes cannot be buffed on a belt polisher. Where these surfaces are finished by hand, they may not match the belt finish.

Buffed finishes should be reserved for narrow, flat surfaces, since application of this process to broad surfaces will result in "oil canning" as a result of the high reflectivity. In addition the belt width limitation will result in overlaps on a broad surface, thereby reducing uniformity of finish and appearance.

Alloys that have an alclad surface may in some instances not be capable of receiving a mechanical finish because the cladding is too thin to withstand such an operation.

A sample of the mechanical finish desired should be examined and the cost and suitability verified prior to making a commitment on the mechanical finish to be used. In addition, the mechanical finish should receive the proposed chemical finish and anodic treatment so that the ultimate finished product can be visualized.

Chemical Finishes

Chemical treatments serve three principal purposes: (1) to clean the aluminum surface in preparation for subsequent finishes; (2) to provide surfaces of uniform electrochemical reactivity to receive anodic coatings; and (3) to etch the surface to obtain specific light reflectance characteristics.

From the standpoint of equipment used and processing time required, chemical treatments are among the least expensive finishing procedures. Mechanical finishing usually involves several operations, with the work performed on individual components or assemblies. Chemical treatment is performed in batches; the finished products are loaded into chemical baths that are part of the anodizing line and sequence of operations.

Several types of chemical treatments are used, and these are classified as: (1) general cleaning treatments (nonetched); (2) etched treatment, which results in a matte finish (most popular of architectural finishes, especially on broad flat surfaces); (3) bright finishes ranging from mirror bright to diffuse bright; and (4) chemical conversion coatings (used for paints, coatings, and laminations, but not to receive anodic finishes). Table 5-6 shows the Aluminum Association classifications of the various chemical treatments.

Since most chemical finishes are achieved by immersing the assembled product in a bath which is usually one of the baths in an anodizing line, the size of the tank may often be the limiting factor in the design of the size of the units to be treated, such as column covers, spandrels, or window frames.

When specifying chemical treatments, care must be exercised that the texture to be obtained is harmonious. For example, because of differing metallurgical characteristics of sheet, extrusions, and castings, the same alloy in these differing forms may not have exactly matching textures following similar chemical treatment. The texture-matching quality may also not be quite the same when similar alloys having different tempers are chemically treated by the same

process. It is recommended that samples be obtained to ensure texture-matching quality whenever various forms, tempers, and alloys are used in the same configuration or in close proximity to one another.

Anodic Finishes

Definition

Anodizing is an electrolytic process in which the aluminum to be anodized is immersed in a specific acid solution through which a direct electric current is passed between the aluminum and the solution, with the aluminum acting as the anode. This causes negatively charged oxygen anions to combine chemically with the aluminum, forming an aluminum oxide film. This electrolytic process results in the controlled formation of a relatively thin but durable anodic coating on the surface of metal, as compared to natural weathering of aluminum, which results in a very thin, soft, and sometimes chalky oxide coating.

Types of Anodizing Processes

While there are several anodizing processes that result in aluminum oxide formations, there are essentially three processes used in architectural applications.

1. *Sulfuric Acid Anodizing:* This process is the oldest and one of the most widely used anodizing processes. In this process sulfuric acid is used as the electrolytic solution. The most notable finishes are the typically clear, transparent coatings. With the advent of newer processes, the sulfuric acid is now often referred to as the "conventional" process. While coloring dyes and pigments are also available with this method, it has limited architectural applications. At present only gold-color mineral pigments are suitable for colorfast exterior applications. As noted in Table 5-7, anodic coatings A31 and A41 are typical clear coatings of this sulfuric acid process. A33 and A43 represent the impregnated colors. Integral colors noted in Table 5-7 as A42 may also be obtained with the conventional method using certain controlled alloys to produce color. If the hardcoat integral color is desired, it must be so specified.

2. *Integral Color Hardcoat Anodizing:* In this process, special patented acid electrolytes are used

TABLE 5-6
CHEMICAL FINISHES (C) ON ALUMINUM

Nonetched Cleaned	
C10	Unspecified
C11	Degreased
C12	Inhibited chemical cleaned
C1X	Other
Etched	
C20	Unspecified
C21	Fine matte
C22	Matte
C23	Coarse matte
C2X	Other
Brightened	
C30	Unspecified
C31	Highly specular
C32	Diffuse bright
C3X	Other
Chemical Conversion Coating[a]	
C40	Unspecified
C41	Acid chromate-fluoride
C42	Acid chromate-fluoride phosphate
C43	Alkaline chromate
C4X	Other

[a]Chemical conversion coatings are shown as part of the chemical finishes. However, they are not a pretreatment for anodic coatings. They result in gray or green colors but are intended as a preparation of the surface to receive organic paint coatings.

TABLE 5-7
ARCHITECTURAL ANODIC FINISHES (A)

Designation	Description	Methods of Finishing
Architectural Class II Finishes (0.4 to 0.7 mil)		
A31	Clear Coating	15% H_2SO_4, 70°F at 12 amps/sq ft, 30 min
A32	Coating with integral color	Color dependent on alloy and anodic process
A33	Coating with impregnated color	15% H_2SO_4, 70°F at 12 amps/sq ft, 30 min followed by dyeing with organic or inorganic colors
A34	Coating with electrolytically deposited color	15% H_2SO_4, 70°F at 12 amps/sq ft, 30 min followed by electrolytic deposition of inorganic pigment in the coating
Architectural Class I Finishes (0.7 mil and over)		
A41	Clear coating	Same as Class II except 60 min electrolysis
A42[a]	Coating with integral color	Color dependent on alloy and anodic processes
A43	Coating with impregnated color	Same as Class II except 60 min of electrolysis
A44	Coating with electrolytically deposited color	Same as Class II except 60 min of electrolysis

[a]Since the A42 finish is obtained by either the conventional or the hardcoat process, the type of process must be specified.

and the integral color is obtained by means of the electrolyte, the aluminum alloy, higher electric current densities, higher voltages, and more accurate control of the electrolyte composition and temperature. The oxide coatings developed through this method are more dense and have higher resistance to abrasion than the "conventional" anodic coatings. The colors range from the light bronzes to medium and dark bronze, and from light grays to dark grays and black. As noted in Table 5-7, anodic coatings A32 and A42 are typical coatings derived by this process.

3. *Electrolytically Deposited Colors:* A proprietary method known as "Analok," licensed by the Aluminum Company of Canada Ltd., produces a range of lightfast colors similar to those produced by the integral color hardcoat process. In this method, noted in Table 5-7 as A34 and A44, stable metallic compounds are electrolytically deposited at the base of the pores in a previously formed oxide coating.

Designation of Anodic Finishes

Table 5-7 shows a modified Aluminum Association designation system for anodic finishes that are used primarily for architectural purposes: Class I and Class II.

Architectural Class I Finishes are those more than 0.7 mil thick and weighing more than 27 mg/sq in. They are appropriate for interior use where they are subject to normal wear and for exterior use where they are subject to the effects of weathering.

Architectural Class II Finishes are those with thicknesses of between 0.4 and 0.7 mils with corresponding weights of from 17 to 27 mg/sq in. They are utilized for interior items that are not subjected to excessive wear or abrasion and for exterior surfaces that are within easy reach and can be maintained on a regular basis.

Sealing the Anodic Coating

The final step in the anodizing process, and one that must be specified, is the sealing of the porous oxide coating. The coating has literally billions of microscopic pores, and these must be closed to prevent absorption of contaminents that will stain and corrode the surface. Both clear and integrally colored anodic coatings are generally sealed in pure or deionized boiling water to seal the pores and prevent further absorption, as shown in Figure 5-1.

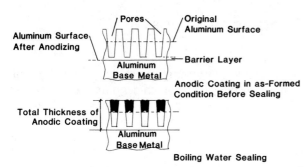

FIGURE 5-1 Sealing mechanism of anodic coatings. (Courtesy of Aluminum Company of America.)

Standard Finish Designations

The Aluminum Association Designation System for aluminum finishes involving clear and color-anodized coatings consists of a letter designating the mechanical treatment (M), the chemical treatment (C), and anodic (A) finishes; and two-digit numbers indicating the specific finish, all of which are shown in Tables 5-5, 5-6, and 5-7.

Examples of how the system is used are illustrated as follows. The designation is preceded by the letters AA to identify it as an Aluminum Association designation.

AA-M21   When this designation is specified, a smooth specular finish is obtained by mechanical treatment.

AA-M32 C12 A31   When this designation is specified, a medium satin finish (M32), an inhibited chemical cleaning (C12), and a clear anodic coating using a sulfuric acid bath for 30 minutes (A31) is obtained.

Quality Control of Anodic Finishes

There are three basic qualities of the anodic coating that must be ensured and these are: (1) the coating weight; (2) the coating thickness; and (3) the resistance of the coating to staining. These qualities can be ascertained by testing of random samples.

Coating Weight: ASTM B137 is a method of test for coating weight which is determined by stripping the coating in phosphoric-chromic acid solution and weighing the sample prior to and after the test procedure.

Coating Thickness: ASTM B244 is a method of test for coating thickness which is determined by the lift-off effect of a probe coil that contacts the coating and generates eddy currents in the metal substate.

Resistance to Staining: This property is obtained by proper sealing of the anodic coating described previously and is ascertained by one of the following test methods:

ASTM B136   Standard Method for Measurement of Stain Resistance of Anodic Coatings on Aluminum

ASTM B457   Standard Method for Measurement of Impedance of Anodic Coatings on Aluminum

International Standard ISO-2931   Assessment of Quality of Sealed Anodic Oxide Coatings by Measurement of Admittance of Impedance. A minimum impedance of 100 kilohms is required.

International Standard ISO-3210   Assessment of Sealing Quality by Measurement of the Loss of Mass after Immersion in Phosphoric-Chromic Acid Solution. The maximum dissolution weight loss should be no more than 2.6 mg/sq in.

Color and Color Matching of Color-Anodized Aluminum

The color and degree of color matching of color-anodized aluminum are of the utmost concern to the architect. The problem concerns not only the architect's design selections but also the architect's specifications and the agreed-upon method of ascertaining the degree of color acceptability and color matching between the architect and the finisher.

When designing a structure, the architect should break up the panel design, using projections or reveals at panel joints that reduce or eliminate undesirable appearance because of slight variations in color of adjacent panels. The architect must also recognize that it is difficult to color-match extrusions and sheet and may want to accentuate the differences.

The selection of matte chemical finishes and bright chemical finishes likewise have an effect on color. Matte finishes will result in a lighter apparent color, and bright finishes will result in a darker apparent color. These apparent color variations result from the reflectivity of the metal substrate as well as from the anodic coating. Only controlled alloys should be specified for color-anodized finishes.

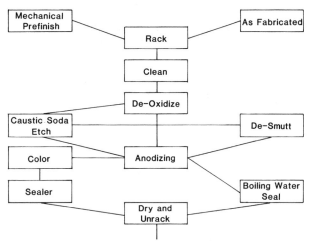

FIGURE 5-2 Steps in the anodizing process. (Courtesy of Aluminum Company of America.)

As an additional control, representative samples should be obtained from a finisher or the major producers. Two samples should be selected to establish the light-and-dark range of colors that will be permitted on the project. During erection, the extreme colors should not be placed adjacent to one another but kept separated by an intermediate color.

During production, an inspection procedure can be specified that is mutually agreeable to the architect and finisher. This may include visual comparison of previously approved color range standards under certain indoor lighting arrangements or outdoor inspection in natural light.

Steps in the Anodizing Process

Since anodic coatings reflect the surfaces under them, the pretreatment of these surfaces by mechanical, chemical, and electrochemical techniques is essential to the attainment of a good anodic coating. The basic steps in the anodizing line after mechanical and chemical pretreatment are shown in Figure 5-2.

APPLIED COATINGS

In addition to the electrochemical anodic finishes, applied coatings may be used as a covering material for aluminum. These coatings are essentially organic, porcelain enamel, or laminated.

Organic Coatings

For architectural applications, only those organic coatings recognized as "high-performance" types with life expectancies of at least 20 years are described here. These consist of the fluorocarbons, the siliconized acrylics and polyesters, and the plastisols. In turn, to qualify as high-performance coatings, standards have been established by the Architectural Aluminum Manufacturers Association (AAMA). AAMA 605.1 "Specification for High Performance Organic Coatings on Architectural Extrusions and Panels" is a specification that describes test procedures and requirements for high-performance pigmented organic coatings.

Fluorocarbons

Fluorocarbon coatings (polyvinylidene fluoride) are based on polymer resins developed by the Pennwalt Corporation and are known as "Kynar 500" resins. The resins are formulated by a number of paint manufacturers into coatings that are highly resistant to UV, weathering, and abrasion.

Siliconized Polymers

The siliconized polymer coatings are based on combinations of high-quality organic polymers and silicones. One type is siliconized acrylic and the other siliconized polyester. They provide good resistance to color change and chalking as well as extended gloss retention.

Plastisols

Plastisols are dispersion coatings based on high molecular weight polyvinyl chloride homopolymer resins dispersed in a plasticizer. They provide good corrosion resistance, weatherability and durability. Plastisols also have good chemical resistance and are used in chemical processing plants.

Porcelain Enamel Coatings

Porcelain enamel coatings are essentially vitreous, inorganic coatings bonded to aluminum by fusion at a temperature of 800°F or higher. Porcelain enamel is an alumina-borosilicate glass mixed with other ingredients to provide color and other desirable characteristics, such as resistance to heat, abrasion, and corrosion.

The alloys used for aluminum sheet, extrusions, and castings have been discussed previously under the heading Aluminum Alloys.

The Porcelain Enamel Institute (PEI) has long been active in promoting quality standards for architectural porcelain enamel and has produced standards for quality control, alloy selection, and test methods. A recommended specification for exterior use of porcelain enamel is PEI: ALS-105.

EXTRUSIONS

One of the advantages of using aluminum for architectural purposes is the comparative ease with which it may be extruded. This allows the designer freedom of choice in fashioning a detail that would involve more intricate metal working with other architectural metals.

When aluminum is extruded, a heated ingot is placed in an extrusion press and forced through a shaped opening. The dies can create a limitless variety of one-piece shapes with external and internal ribs, channel curves, flats, slots, grooves, and oddly shaped openings. The extrusion die is comparatively inexpensive, and this feature too allows the designer to utilize this technique of extrusion more frequently. At present, aluminum extrusions are produced with circle sizes ranging from ¼ inch to 31 inches.

METALLURGICAL JOINING PROCESSES

For a comprehensive discussion on the subject of welding, brazing, and soldering, see *Welding Aluminum*, published by the American Welding Society.

Welding

Aluminum is readily joined by welding, and most welding techniques may be used. When appearance is of primary importance, fusion welding using the gas tungsten-arc and gas metal-arc produce the best as-welded appearance.

Welded parts may also receive chemical pretreatments and electrochemical anodic finishes. Resistance, ultrasonic, and pressure welds are least noticeable after these treatments.

Brazing

Brazing forms joints of excellent appearance that require little or no finishing. When anodizing is required, brazing filler metals darken during the finishing. To avoid this problem, complete flux removal is required prior to anodic treatment.

Soldering

Soldering of aluminum is not the best solution to joining of aluminum. Welding and brazing are preferable for joints requiring good strength. Other techniques that can be used to avoid soldering include adhesive bonding, mechanical fastening, or epoxied joints. Soldering is not recommended for sheet metal joining.

COPPER

Copper and its alloys (principally zinc) are probably the oldest metals used for ornamental metal work in building construction. This is undoubtedly due to the fact that, of all architectural metals, copper and its alloys provide the inherent advantage of integral color. The different alloys have varying colors and, as will be explained later, problems can arise in otherwise well-executed architectural detailing because of failure to select proper alloys for color match and color harmony.

The copper alloys primarily used for ornamental metalwork have been and still are called bronzes.

They are, however, truly brasses except for some of the casting alloys. The wrought sheet, plate, and extrusion copper alloys are essentially alloys of copper and zinc, which is the definition for brass. Only some of the casting alloys containing 2% or more of tin can be classified as bronzes.

PHYSICAL PROPERTIES

When cold worked, copper alloys may exhibit tensile strengths ranging from 36,000 to 70,000 psi for some silicon bronze alloys. Tensile values for red brass 230, Muntz metal 280, and architectural bronze 385 fall in the range of 40,000 to 60,000 psi. These strengths compare favorably with those of aluminum and carbon steel. The modulus of elasticity for copper alloys in tension ranges from 14 million to 17 million.

The thermal coefficient of expansion of copper alloys is approximately 0.00001 per °F. A useful approximation allows ⅛ inch expansion per 10 feet of length for each 100°F temperature change.

Since the cost of copper is high as compared to aluminum and steel, light structural steel core members are used to stiffen an element using sheet copper alloys as a covering.

COPPER AND COPPER ALLOYS

Alloy Designation System

The major copper alloys, their composition, natural color, and final weathered colors are shown in Table 5-8. The final weathered colors are achieved in about six years.

TABLE 5-8
PRINCIPAL COPPER ALLOYS

Alloy No. & Name	Composition	Natural Color	Weathered Colors
110 copper	99.9% Cu	salmon red	reddish brown to gray-green
122 copper	99.9% Cu .02% P	salmon red	reddish brown to gray-green
220 commercial bronze	90% Cu 10% Zn	red gold	brown to gray-green
230 red brass	85% Cu 15% Zn	reddish yellow	chocolate brown to gray-green
260 cartridge brass (yellow brass)	70% Cu 30% Zn	yellow	(not for exterior use)
280 Muntz metal	60% Cu 40% Zn	reddish yellow	red-brown to gray-brown
385 architectural bronze	57% Cu 3% Pb 40% Zn	reddish yellow	russet brown to dark brown
655 silicon bronze	97% Cu 3% Si	reddish old gold	russent brown to mottled dark gray-brown
745 nickel silver	65% Cu 25% Zn 10% Ni	warm silver	gray-brown to mottled gray-green
796 leaded nickel silver	45% Cu 42% Zn 10% Ni 2% Mn 1% Pb	warm silver	gray-brown to mottled gray-green

TABLE 5-9
FORMS, STANDARDS, AND USES OF COPPER ALLOYS

Alloy No.	Forms Available	Standards*	End Uses
110	Strip, sheet	B152	Roofing metal
122	Tube	B75 B88	Roofing metal, doors, lighting fixtures, railings
220	Strip Wire Bar, rod	B36 B134 —	Unlimited usage
230	Strip Pipe Tube	B36 B43 B135	Unlimited use
260	Strip Tube Pipe	B26 B135 —	Interior use, interior-lighting, fixtures, signs
280	Plate, sheet Strip	Fed. Spec. QQ-B-613	Curtain walls, entrance doors
385	Extrusions	B455	Curtain walls, entrance doors
655	Plate, sheet Pipe, tube	B97 B315	Artwork, entrance doors, windows
745	Sheet, strip Bar Rod Tube	B122 B151 B206 —	Artwork, bank equipment, elevator cabs
796	Bar extrusions, rod	—	Artwork, bank equipment, elevator cabs

* Standards are ASTM except when noted Fed. Spec.

Principal Forms, Standards, and Uses

The various copper alloys are available in a variety of forms, such as rods, bars, tubes, and sheets. Table 5-9 illustrates the principal forms that are available, the standards to which they are produced, and typical end uses.

Color Matching of Copper Alloys

Color match and color harmony are prime considerations for the architect when using copper alloys. It is not unusual to find as many as four alloys in combination for some designs to achieve color match and harmony. Table 5-10 shows color matching of various forms of copper alloys as equated to sheet and plate forms.

STAINING AND CORROSION

Sulfur compounds, stemming from combustion of fossil fuels and combined with water in the form of acid rain, are the chief atmospheric corrosive agents of copper and its alloys. The weathering of copper alloys over the years is attributable to this corrosive action. Concentrated attack may result in thinning of the metal to failure. Moderate weathering produces the patinas that enrich the coloring of copper alloys.

Wherever dissimilar metals are in direct contact in the presence of moisture, galvanic corrosion of the less noble metal in the couple may occur (see Chapter 1, Table 1-5). Under such conditions, provision should be made to insulate the materials to avoid corrosion by using paint primers, sealants, or tapes.

Where copper roofing or flashing shed water onto porous stonework such as limestone or marble, the detailing of such installations should be refined to

TABLE 5-10
COLOR MATCHING COPPER ALLOYS[a]

Sheet and Plate Alloys	Extrusions	Castings	Fasteners	Tube & Pipe	Rod & Wire	Filler Metals
Alloy 110 Alloy 122 Copper	Alloy 110 Copper (simple shapes)	Copper (99.9 min.)	Alloy 651 Low-silcon bronze (fair)	Alloy 122 Copper	Alloy 110 Copper	Alloy 189 Copper
Alloy 220 Commercial bronze, 90%	Alloy 314 Leaded commercial bronze	Alloy 834	Alloy 651 Low-silicon bronze	Alloy 220 Commercial bronze, 90%	Alloy 220 Commercial bronze, 90%	Alloy 665 High-silicon bronze
Alloy 230 Red brass, 85%	Alloy 385 Architectural bronze	Alloy 836*	Alloy 651 Low-silicon bronze (fair) Alloy 280 Muntz metal	Alloy 230 Red brass, 85%	Alloy 230 Red brass, 85%	Alloy 655 High-silicon bronze (fair)
Alloy 260 Cartridge brass, 70%	Alloy 260 Cartridge brass, 70% (simple shapes)	Alloys 852, 853	Alloy 260 Cartridge brass, 70% Alloy 360 Alloy 464 Alloy 465	Alloy 260 Cartridge brass, 70%	Alloy 260 Cartridge brass, 70%	Allow 681 Low-fuming bronze (poor)
Alloy 280 Muntz metal	Alloy 385 Architectural bronze	Alloys 855* 857	Alloy 651 Low-silicon bronze (fair) Alloy 280 Muntz metal	Alloy 230 Red brass, 85%	Alloy 280 Muntz metal	Alloy 681 Low-fuming bronze
Alloy 655 High-silicon bronze	Alloy 655 (simple shapes)	Alloy 875	Alloy 651 Low-silicon bronze Alloy 655 High-silicon bronze	Alloy 651 Low-silicon bronze Alloy 655 High-silicon bronze	Alloy 651 Low-silicon bronze Alloy 655 High-silicon bronze	Alloy 655 High-silicon bronze
Alloy 745 Nickel-silver	Alloy 796 Leaded nickel-silver	Alloy 973	Alloy 745 Nickel-silver	Alloy 745 Nickel-silver	Alloy 745 Nickel-silver	Alloy 773 Nickel-silver

Source: Courtesy, Copper Development Association, Inc., Greenwich, CT 06836.

[a]Alloys to be used in various forms, for best color match with certain sheet and plate alloys. Color of surfaces compared after identical grinding or polishing.

prevent such run-off directly onto these surfaces, since it may cause significant disfigurement.

TEMPERING AND HEAT TREATMENT

In the copper alloy family, tensile and yield strengths are increased primarily by means of alloying and cold working. Although heat treating finds some limited use as a strengthening mechanism, it is not applicable to any of the principal architectural copper alloys.

FINISHING

Standard designations for copper alloy finishes are published by the Copper Development Association, Greenwich, Connecticut, in the *Copper, Brass, Bronze Design Handbook*. These consist of mechanical finishes—M; chemical finishes—C; organic coatings—O; and laminated finishes—L. A summary of these standard designations is shown in Table 5-11.

TABLE 5-11
SUMMARY OF STANDARD DESIGNATIONS FOR COPPER ALLOY FINISHES

Mechanical finishes (M)			
As Fabricated	Buffed	Directional Textured	Non-Directional Textured
M10-unspecified mill finish	M20-unspecified	M30-unspecified	M40-unspecified
M11-specular as fabricated	M21-smooth specular*	M31-fine satin*	M41-(unassigned)
M12-matte finish as fabricated	M22-specular*	**M32-medium satin**	**M42-fine matte***
M1x-other (to be specified)	M2x-other (to be specified)	**M33-coarse satin**	M43-medium matte
		M34-hand rubbed	M44-coarse matte
		M35-brushed*	M45-fine shot blast
		M36-uniform	M46-medium shot blast
		M3x-other (to be specified)	M47-coarse shot blast
			M4x-other (to be specified)

Chemical finishes (C)	
Non-Etched Cleaned	Conversion Coatings
C10-unspecified	**C50-ammonium chloride** (patina)
C11-degreased	**C51-cuprous chloride-hydrochloric acid** (patina)
C12-cleaned	**C52-ammonium sulfate** (patina)
C1x-other (to be specified)	C53-carbonate (patina)
	C54-oxide (statuary)
	C55-sulfide* (statuary)
	C56-selenide (statuary)
	C5x-other (to be specified)

Coatings			
Clear Organic (O)	Laminated (L)	Vitreous and Metallic	Oils and Waxes
AIR DRY (gen'l arch'l work)	L90-unspecified	since the use of these finishes in architectural work is rather infrequent, it is recommended that they be specified in full, rather than being identified by number.	these applied coatings are primarily used for maintenance purposes on site. Because of the broad range of materials in common use, it is recommended that, where desired, such coatings be specified in full.
O60-unspecified	L91-clear polyvinyl fluoride		
O6x-other (to be specifed)	L9x-other (to be specified)		
THERMOSET (hardware)			
O70-unspecified			
O7x-other (to be specified)			
CHEMICAL CURE			
O80-unspecified			
O8x-other (to be specified)			

Source: Courtesy, Copper Development Association Inc., Greenwich, CT 06836.

ᵃIn this listing, those finishes which are printed in boldface type are the ones most frequently used for general architectural work; those marked * are commonly used for hardware items.

Chemical Coloring

Some of the chemical finishes hasten the natural weathered effect that generally results from exposure to the weather. The verde antique finishes are developed using acid chloride treatment or acid sulfate treatments.

Statuary (oxidized) finishes are produced in light, medium, and dark colors, depending upon both the concentration and the number of applications of the chemical coloring solutions. Two to ten percent aqueous solutions of ammonium sulfide, potassium sulfide (liver of sulfur) or sodium sulfide (liquid sulfur) are used to produce statuary finishes.

DESIGN CRITERIA

Sheet and Strip

When designing in sheet, it is necessary to avoid "oil canning" or buckling. With thinner gages, distortion becomes more pronounced. Broad, bright, reflective surfaces accentuate minor variations in flatness and become apparent. Distortion can be reduced by increasing the gage or by using a backing material to increase rigidity. Natural weathering, chemical coloring, and texturing tend to mask minor distortions by reducing surface reflectivity.

Unsupported broad, flat areas such as spandrels or column covers usually require a metal thickness greater than 0.100 inch to avoid "oil canning." By introducing patterns or textures the increased rigidity may permit gage reductions to the range of 0.064 to 0.100 inch. These reduced thicknesses are also suitable for metals laminated or bonded to suitable substrates.

Extrusions

As a general working guide, extruded shapes are limited in cross section to shapes having a greatest diagonal dimension which can be contained within the perimeter of a 6-inch circumscribed circle and as shown in Table 5-12.

TABLE 5-12
EXTRUSION DESIGN CRITERIA

Diameter of circumscribing circle (inches)	Minimum gage of extrusion (inches)
2	.064
3	.080
4	.093
5	.108
6	.125

FABRICATING PROCESSES

Forming

Architectural copper alloys lend themselves to a variety of forming processes as shown in Table 5-13.

Joining

Fasteners

The use of fasteners (screws, bolts, rivets) to join architectural copper alloys is the simplest and most common joining method. Color matching of fastener to metal alloy is important and is shown in Table 5-14.

Adhesive Bonding

Copper alloys can be bonded to steel, plywood, hardboard, and similar substrates. As a result, thinner gages can be used, resulting in cost reductions and in panels having flat, broad surfaces without "oil canning." Alloy 110 can be used in such applications in gages ranging from 0.002 to 0.032 inch. Muntz metal 280 can be applied in gages as low as 0.032 inch.

For exterior applications, thermosetting or high-quality thermoplastic adhesives are used. Since moisture may enter at edges, edge design and detailing are crucial to prevent peeling and moisture entry.

Adhesive Joining

Railings, intersecting members, and similar details can be joined by the use of bonding adhesives and sleeved connections with concealed pins and fasteners.

Metallurgical Joining Processes

Copper alloys employ brazing, soldering, and welding in joining metal parts. Brazing is the preferred method, in terms of adequate joint strength and soundness. However, blind or concealed joints are preferred with brazing, since color matching is fair to poor.

TABLE 5-13
ALLOYS SUITABLE FOR FABRICATING PROCESSES

Alloy	Bending	Brake Forming	Extrusion	Cold Forging	Hot Forging	Roll-Forming	Spinning
110	X			X	X	X	X
122	X						
220	X	X		X	X	X	X
230	X	X		X	X	X	X
260	X	X		X	X	X	X
280	X	X			X		
385			X		X		
655	X	X		X	X	X	
745	X	X		X		X	X
796			X				

Soldering is used primarily for roofing and flashing to seal joints. These joints ae mechanically reinforced to ensure adequate strength. Soldering should be confined to concealed joints, since solders do not color match with the copper alloys.

Welding of copper alloys should be restricted to the silicon bronzes that are readily welded and produce good color matches and sound welds.

STAINLESS STEEL

Stainless steels make up a family of corrosion- and rust-resistant iron-base alloys containing a minimum

TABLE 5-14
MECHANICAL FASTENERS

Fastener Alloy	Color Match
260	Color matches alloy 260
280	Matches alloy 385
360	Matches alloy 260
464 and 467	Fair color match with alloy 385
485	Fair to good color match with alloys 280 and 385
745	Color matches with alloy 745 and 796

of 12% chromium. The corrosion resistance is improved by adding more chromium. Nickel and manganese when alloyed with the chromium-iron-base metal produces special characteristics such as strength, toughness, and ease of fabrication.

Stainless steel is the strongest, most durable, and most corrosion resistant of all the architectural metals. It is likewise nonstaining and can therefore be used with other materials such as stone, metals, and clay tile products without the danger of staining or deterioration.

Because of its strength, lighter gages can be used as compared to other metals, thereby affecting economies. Also, because of its corrosion-resistant properties, minimal maintenance is required of stainless steel.

ALLOY CLASSIFICATION

Classification

While there are over 40 different stainless steel alloy compositions, only seven are commonly used in architectural metalwork. In addition, stainless steels are divided into three groups according to metallurgical structure: austenitic, martensitic, and ferritic. While this classification is of interest chiefly to metallurgists, there are basic differences that are of import to architects.

1. *Austenitic Stainless Steels* (AISI 200 and 300 Series) contain nickel and are essentially nonmagnetic. They are hardened by cold working but not by heating. They are ductile and can be fabricated and welded easily. The types used primarily in architectural metalwork are 201, 202, 301, 302, 304, and 316.

2. *Ferritic Stainless Steels* (AISI 400) contain chromium as the primary alloying element and are magnetic. They are hardened only slightly by heat treating and can be hardened moderately by cold working. The type used primarily in architectural metalwork is 430.

3. *Martensitic Stainless Steels* contain a magnetic alloy with a limited application in architectural uses.

Types

The various types of stainless steel are usually employed as follows.

For sheet, strip, plate, bars, tubing, and extrusions that find application in both exterior and interior architectural metalwork such as column covers, doors, fascias, mullions, panels, windows, and trim, and for roofing metalwork:

TYPE 301	Contains less chromium and nickel than types 302/304 and has slightly higher work-hardening characteristics; also slightly higher tensile strength than 302/304
TYPE 302/304	The most popular and most widely used alloys architecturally, referred to as "18–8" stainless (18% Cr–8%Ni); often used interchangeably
TYPES 201 and 202	Lower in nickel content and higher in manganese content than the 302/304 types but a little more difficult to form

For applications at or near marine or saltwater atmospheres and in industrial applications where corrosive conditions obtain:

TYPE 316	A modification of types 302/304 with a higher nickel content and the addition of

2 to 3% molybdenum that adds to corrosion resistance and pitting resistance.

For interior architectural applications where less resistance to corrosion is tolerable:

TYPE 430	A ferritic chrominum stainless steel alloy with no nickel and less costly than austenitic types described above.

For fasteners in either exposed or protected locations:

TYPE 305	Used for bolts, nuts, screws, and other fasteners, compatible with types 302/304 insofar as appearance, corrosion and durability are concerned

For fasteners in locations protected from weather:

TYPE 410	used for mechanical fasteners where less corrosion resistance is required

Physical Properties

Yield strength of stainless steel alloys is between 30,000 and 50,000 psi; tensile strength between 75,000 and 115,000 psi; and modulus of elasticity about 28,000,000 psi. The coefficient of thermal expansion is 0.0000094.

FINISHES

Stainless steel sheet and strip are supplied from the mill in various standard finishes. Polished finishes (Nos. 3, 4, 6, 7, and 8) can be applied to plate, bars, tubular products, and extrusions. Table 5-15 shows the extent of the standard mill finishes.

Proprietary Finishes

In addition to the standard mill finishes, many proprietary finishes are available from individual producers. It is best to obtain samples and explanations as to cost, availability, and efficacy of these proprietary finishes on specific shapes and forms.

TABLE 5-15
STANDARD MILL FINISHES

Type	Description
Sheet-Rolled	
No. 1	A dull finish, nonarchitectural
No. 2D	A dull, nonreflective finish
No. 2B	A bright, moderately reflective finish
Bright annealed	A bright, highly reflective finish
Sheet-Polished	
No. 3	An intermediate polished finish
No. 4	A bright machine-polished finish with a visible "grain", the finish most frequently used for architectural applications
No. 6	A dull satin finish
No. 7	A bright, highly reflective finish
No. 8	A bright "mirror" finish
Stainless Steel Strip	
No. 1	Similar to No. 2D for sheet
No. 2	Similar to No. 2B for sheet
Bright annealed	Similar to "bright annealed" for sheet

Joining

Stainless steel can be assembled or joined by metallurgical means (welding, and soldering) and by mechanical fasteners (bolts, screws, rivets).

Welding of stainless steel is readily performed by all the common fusion and resistance methods. In most instances, grinding and polishing of welds removes any trace of the weld, which blends with the adjacent surface.

Soldering is accomplished with phosphoric fluxes that will not corrode stainless steel. Soldering should be used to seal or fill a joint, but not to provide structural strength.

Mechanical fastening of all-stainless-steel assemblies should be accomplished with stainless steel screws, bolts, and rivets, to ensure a permanent noncorroding attachment.

Adhesive bonding of stainless steel to plywood and hardboard in the production of veneer panels and sandwich panels is possible with the development of structural adhesives such as epoxies and polyurethanes.

FABRICATING PROCESSES

Forming

Stainless steels for architectural applications may be cut by shearing and sawing. They may be blanked, nibbled, punched, and drilled; also brake formed, roll formed, cast, and extruded.

Castings are used for hardware and for ornamental purposes in sculpture and plaques.

Extrusions of stainless steel are likewise available in angles, railings, stiles, thresholds, and other shapes. There are limitations on sizes based on the extrusion technique. Since extrusion is a constantly evolving procedure, designers should consult the extruder or fabricator to obtain information on the latest capabilities in the extrusion process.

The high ductility of stainless steel permits the use of sharp bends and moderately deep stamped patterns. Sheets of 18 gage or thinner can be bent 180° and flattened without cracking.

STEEL

Steel, an alloy of iron, is used by architects not only as an integral part of structural systems in buildings but, when exposed, as part of the architectural ornamental expression. Exposed structural systems used architecturally include space frames and arches to span large, unobstructed floor areas such as convention halls, exhibition spaces, or sports arenas. Others include cable-supported structures, and still others include low buildings where fireproofing is not mandated by building codes and the steel frame may be left exposed.

Steel is also used for ornamental metalwork to enhance the appearance of buildings. Some structures are famed for their beautiful grille work, stairs, railings, sculpture and ornamental metalwork.

The classification and identification of steel is rather complex and involves, among other things: (1) chemical composition; (2) method of manufacture; (3) mechanical properties; (4) heat treatment; and (5) reference to a recognized standard.

For architectural purposes, carbon steel and high-strength, low-alloy steel (HSLA or "weathering steel") are the primary steels used to produce such effects. Stainless steel is a steel alloy but for purposes of singular identification as an ornamental metal it is discussed separately in this chapter.

CARBON STEEL

Classification

Steels are classified as carbon steels when: (1) no minimum content is specified or required for the alloying elements; (2) the amount of carbon does not exceed 2%; and (3) the amount of iron exceeds 95%.

Carbon steels are likewise classified by grade on the basis of their carbon content. Amounts of up to 0.8% carbon increase the strength and hardness of carbon steel and decrease the ductility. Table 5-16 shows types of carbon steel based on the percentage of carbon.

Physical Properties

Carbon steel used for architectural purposes includes mild steel, medium steel, and very mild steel. Of these, the mild steels are the most widely used and are generally carried in stock by warehouses and fabricators. Mild steels have a tensile strength in the range of 36,000 to 65,000 psi, form easily, and retain true sections when fabricated.

Shapes

Bar sizes of carbon steel include channels, angles, Tees, Zees, and rolled sections having a maximum dimension of the cross-section of less than 3 inches.

Structural size shapes are rolled, flanged sections having at least one dimension of the cross-section 3 inches or greater.

Most bar size shapes and structural size shapes of carbon steel are covered by ASTM A36 for chemical requirements, tensile properties, and bend test requirements.

Pipe and Tubing

Pipe and tubing as defined earlier in this chapter under Standard Stock Mill Products are available for a variety of architectural design purposes. They may be obtained in standard weight, extra-strong, and double extra-strong.

Several ASTM standards are available by which to specify pipe and tubing as follows:

ASTM A53   Welded and Seamless Steel Pipe; covers black and hot-dipped galvanized products

ASTM A500   Cold Formed Welded and Seamless Carbon Steel Structural Tubing in Rounds and Shapes

ASTM A501   Hot-Formed Welded and Seamless Carbon Steel Structural Tubing

TABLE 5-16
CARBON STEEL TYPES

Type	Carbon Steel Content	Description
Very mild	0.05–0.15	An extra-soft steel, tough, ductile and used for sheets, wire, pipe, and fastenings.
Mild	0.15–0.25	Strong, ductile, machinable, used for buildings, bridges, boilers.
Medium	0.25–0.35	Harder and stronger than mild steel, used for machinery, shipbuilding, and general structural purposes.
Medium-hard	0.35–0.65	Used where wear and abrasion is encountered.
Spring	0.85–1.05	Used for spring manufacture.
Tool	1.05–1.20	Hardest and strongest grade.

FIGURE 5-3 East Wing, National Gallery of Art, Washington, DC: interior view illustrating space frame. I.M. Pei & Partners, Architects. (Photographer, Ezra Stoller © ESTO.)

FIGURE 5-4 John F. Kennedy Library, Boston, MA: interior view illustrating space frame. I.M. Pei & Partners, Architects. (© Mona Zamdmer 1979, Photographer.)

ASTM A512   Cold-Drawn Buttweld Carbon Steel Mechanical Tubing

ASTM A513   Electric Resistance-Welded Carbon and Alloy Steel Mechanical Tubing

ASTM A618   Structural Tubing, Hot Formed, Welded and Seamless

See Figures 5-3 and 5-4 for illustrations of pipe and tubing used for space frames.

Sheet and Strip

Sheet and strip are used architecturally for components of buildings such as curtain walls and fascias; roofs; for interior walls and partitions; doors and windows; floors and ceiling systems; and signage.

Steel sheet and strip are made up of carbon steels; high-strength, low-alloy steels; and full alloy steels.

Hot-rolled sheet and strip are produced by squeezing hot steel ingots with huge rolls repeatedly until the desired thickness is reached. Cold-rolled sheets are formed by further rolling hot-rolled sheets after they have been allowed to cool and have been pickled to remove scale.

The most popular steel sheet and strip for architectural purposes are as follows:

Hot-rolled carbon steel
Cold-rolled carbon steel
High-strength, low-alloy weathering steel
Sheet for porcelain enameling
Zinc-coated steel

These can be further identified by reference to ASTM standards as follows:

ASTM A570   Sheet and Strip, Hot Rolled, Structural Quality

ASTM A611   Sheet and Strip, Cold Rolled Structural
                        Quality
ASTM A424   Sheet for Porcelain Enameling
ASTM A446   Sheet, Hot Dipped Galvanized Struc-
                        tural Quality
ASTM A591   Sheet, Electrolytic Zinc Coated
ASTM A606   Sheet and Strip, Hot Rolled and Cold
                        Rolled, Improved Atmospheric Corro-
                        sion Resistance

Metallic Coated Steel Sheet

Aluminum: Aluminum-coated steel sheet, Type
II, consists of steel sheets, factory finished on both
sides, with a coating of commercially pure aluminum.
These sheets combine the strength of steel with the
corrosion resistance of aluminum and are used for
roofing, siding, and other building products.

Terne Metal: A mixture of 85% lead and 15% tin
used as a coating on both sides of steel sheet. Coating
weights are typically 20 pounds and 40 pounds. Used
primarily for roofing, terne metal has long-term permanence when painted and maintained. The standard
for terne-coated steel sheet is ASTM A308.

Other metallic coatings: Steel sheet is frequently
plated with other metals such as brass, copper,
chrominum, or tin, to combine the qualities of the
coating (such as color and texture) with the strength
of steel and a reduction in cost.

Fabrication

Carbon steels can be hot rolled, cold rolled, forged,
cast, welded, punched, blanked, and cut, but not extruded.

Hot-rolled steel has a thin, tight scale over a
smooth-rolled surface; corners are usually slightly
rounded.

Cold-rolled steel is produced by passing cleaned
or pickled metal between heavy rolls or dies to work
the metal while it is cold. The process hardens and
stiffens the steel, increases its tensile strength, and improves the surface. Cold-rolled or cold-finished products are more accurate as to size and gage than hot
rolled products and are free from scale.

HIGH-STRENGTH, LOW-ALLOY STEEL (WEATHERING STEEL)

Weathering steel is a high-strength, low-alloy steel that
is remarkably resistant to corrosion from normal atmospheric exposure and attains a tight oxide coating
of varying colors that are architecturally pleasing. The
color of weathered steel can be described as rus-
set brown with an intermixing of browns, reds, and
blues.

In ordinary carbon steel, rust begets rust. The corrosion resistance of high-strength, low-alloy weathering steel depends upon the formation of a dense,
stable layer of rust that prevents oxygen and other
contaminants from continued reaction with the base
metal. The corrosion resistance of weathering steel is
approximately 4 times that of carbon steel.

The action of continued wetting and drying of the
surface is essential to develop this tight oxide covering
over a period of about 3 years. An indoor environment
or an arid climate are not conducive to the full development of this oxide coating.

Careful design is essential in the detailing of various
building elements utilizing weathering steel. Surfaces
must drain properly to permit drying; otherwise, normal, loose rusting occurs. Water that drains or drips
from the steel must not be allowed to impinge on other
materials that will show rust stains or streaking. Porous
materials are vulnerable to such absorption and discoloration.

In designing with weathering steel for the first time,
it is essential that the architect allow the producer to
review the details of design to ensure that the components weather properly and that adjacent materials
are not subject to water run-off from the weathering
steel.

To obtain a uniform weathering that is pleasing in
appearance, mill scale must be removed by cleaning,
sanding, scraping, or wire brushing. Blast cleaning is
perhaps the best preparation. Oil, grease, and chalk
marks must also be removed, or else these areas will
not weather uniformly.

Weathering steel structural shapes and sheet metal
that may be specified to ASTM standards are as follows:

ASTM A242   High-Strength Low-Alloy Structural
                       Steel

FIGURE 5-6 U.S. Steel Building, Pittsburgh, PA: example of exposed weathering steel facade. Harrison & Abramovitz & Abbe, Architects. (Courtesy of U.S. Steel Corp.) This illustration also appears in the color section.

One of the earliest examples of a unique design employing weathering steel is the John Deere Company administration building in Moline, Illinois, designed by Eero Saarinen. See Figures 5-5 and 5-6 for typical examples of buildings utilizing weathering steel.

FIRE SAFETY OF STEEL STRUCTURES

FIGURE 5-5 Ruan Center, Des Moines, IA: example of an unpainted exterior plate wall using weathering steel. Kendall, Griffith, Russel, Artiaga, Architects-Engineers. (Courtesy of U.S. Steel Corp.) This illustration also appears in the color section.

ASTM A588   High-Strength Low-Alloy Structural
            Steel
ASTM A606   Steel Sheet and Strip, Hot Rolled and
            Cold Rolled, High Strength, Low Alloy

Fasteners of high-strength structural bolts that are made of these alloys should be used so that uniform weathering will result. Welding should be performed with compatible welding electrodes.

Elements of weathering steel that are not boldly exposed to the atmosphere—such as glass rebates in lights or such other connections, including contact surfaces of overlapping steel—should be painted to prevent formation of loose rust.

Unprotected steel at temperatures above 1000 F° loses about half its ultimate strength. As a result, steel framing in most structures is encased in concrete, masonry, or gypsum board, or covered with cementitious and fibrous sprayed-on fireproofing or intumescent coatings.

Limiting the temperatures that exposed steel structures may reach during a fire requires designing the system in such a manner as to preclude this temperature rise. One example is the U.S. Steel building in Pittsburgh, PA. The building has liquid-filled columns to circulate water heated by fire away from the affected columns and replace it with cooler water from storage tanks or city water supplies (see Figure 5-7) Another example is the use of flame shields that deflect heat and flames from a burning building away from exterior structural steel spandrels. This design is used in the One Liberty Street building in New York City, as shown in Figure 5-8.

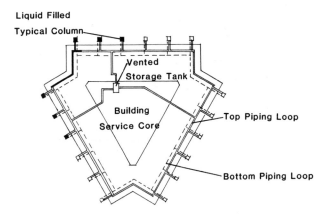

FIGURE 5-7 U.S. Steel Building. Schematic plan of fire-proofing system including liquid-filled exterior columns. (Courtesy of U.S. Steel Corp.)

A new design aid for exposed steel introduced by the American Iron and Steel Institute is *FS-3, Fire Safe Structural Steel.* It is a calculation procedure intended for exterior spandrel beams and exterior columns which may be protected from a building fire by geometry of space and distance. While this method is still in its infancy, it is a rational approach and may open the way for designing some structures utilizing exposed framing that is fire protected by space and geometry.

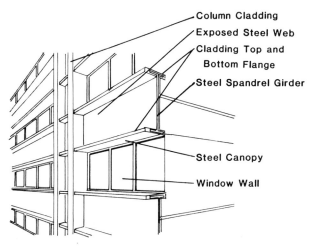

FIGURE 5-8 One Liberty Plaza, New York, NY. Cladding attached to top and bottom flanges form a fire canopy, protecting the exposed steel web of the girder. (See also Figure 8-10.)

METALLIC PROTECTIVE COATINGS

General

Carbon steels are often coated with metallic protective coatings to inhibit corrosion of the base metal. To ensure sound protective coatings, they must be of uniform density and free of pinholes or other discontinuities.

Zinc Coatings

Zinc coatings protect steel based on the galvanic reaction between zinc and steel which causes the zinc to corrode in favor of the steel. See Chapter 1, Compatibility. Zinc is applied to steel products by several different methods as follows:

Hot-Dip Galvanizing: A process of coating steel products by immersing them in a bath of molten zinc after cleaning them. This process provides the surface with a tightly adhering coat of zinc, which is one of the most effective agents in protecting steel from rust. Several ASTM standards are available for hot-dip galvanizing procedures as follows:

ASTM A123   On steel fabricated products, (rolled/pressed/forged shapes)

ASTM A153   On iron/steel hardware

ASTM A386   On assembled steel products

Several classes of hot-dip zinc coatings are available relating to the weight of coating per square foot of surface. The weight of coating to be used is proportional to the severity of the corrosion potential to be expected.

Electrogalvanizing: A process produced by an electric current. By immersing a steel product in the electroplating solution of zinc sulfate or zinc cyanide, a pure zinc coating is deposited whose thickness can be controlled. Heavy coatings such as those provided by the hot-dip method cannot be obtained in the electrolytic process.

Sherardizing: A zinc cementation process wherein the steel product to be coated is surrounded by zinc dust and then heated in an oven. A thin zinc coating is produced over the steel product. This process is limited to small products of complex shape.

Spraying: A process whereby zinc is fed in the form of a wire into a spray gun, where it is melted and projected by air pressure in a hot, atomized spray against the object to be coated.

Painting: Application of a zinc-rich paint on prepared surfaces of steel. Zinc-rich paint is produced to meet the requirements of Steel Structures Painting Council SSPC-PS 12.

ARCHITECTURAL WOODWORK

This chapter is intended and limited to discuss only wood and wood products that are utilized in architectural woodwork. For this purpose architectural woodwork includes millwork, cabinetwork, paneling, and structural glued laminated timber and decking.

The subject matter contained in this chapter is confined to those materials such as solid wood, plywood, particle board, and plastic laminate, which in turn are generally used to produce end products that result in architectural woodwork.

STANDARDS AND REQUIREMENTS

TABLE 6-1
% AVERAGE MOISTURE CONTENT

Dry Southwestern States	Damp Southern Coastal States	Remaining States in Continental United States
6	11	8

CLASSIFICATION OF WOOD

Trees that provide lumber for architectural woodwork are divided as a matter of convenience into two groups—softwoods and hardwoods. The softwoods, in general, are the coniferous or cone-bearing evergreen trees such as the pines, hemlocks, firs, spruces, and cedars. The hardwoods are the deciduous or broad-leafed trees such as the maples, oaks, and birches.

The terms hardwood and softwood refer primarily to the above breakdown of groups and not to the fact that one group is hard and the other is soft.

The softwoods are more commonly used for framing purposes, such as studs, joists, rafters, and posts. The hardwoods are primarily used for interior finishes, flooring, paneling, cabinetry, and furniture where natural finishes are desired.

MOISTURE CONTENT AND SEASONING

Moisture content is defined as the weight of water contained in wood, expressed as a percentage of the weight of the oven dry wood. As wood dries and its moisture content decreases, wood shrinks; conversely, as wood absorbs moisture, it swells.

When equilibrium for moisture content is reached for a condition where the wood will be in service, its tendency to shrink, expand, or warp will be diminished. However this condition never remains constant in actual service because of normal changes in atmospheric moisture. It is therefore important that an approximate equilibrium moisture content be reached.

It is essential that lumber be seasoned until the moisture content is similar to the conditions under which the wood will be used. Lumber is seasoned or dried by natural air drying or by kiln drying. Air drying takes place outdoors where wood is stacked and allowed to season over a period of about two months. Kiln drying takes place in a large "oven" where the rate of seasoning of lumber is controlled and requires only two or three days to accomplish.

The Wood Handbook, published by the Forest Products Laboratory of the Department of Agriculture, recommends the moisture content values for interior woodwork as shown in Table 6-1.

LUMBER GRADES

Grading of lumber refers to its quality. Since wood is a natural material, uniformity is hard to come by. The term grading is best summed up in the U.S. Department of Commerce Product Standard PS-20, American Softwood Lumber Standard, as follows:

The grading for lumber cannot be considered an exact science because it is based on either a visual inspection of each piece and the judgement of the grader or on results of a method of mechanically determining the strength characteristics of structural lumber.

To indicate the grade, the lumber manufacturing associations of softwood lumber stamp the grade on each piece.

Softwood lumber grades are based on PS-20 (identified above) which establishes voluntary guidelines for the softwood industry associations. Each softwood association classifies softwood lumber according to size, use, and manufacturing.

Hardwood lumber grades are established by the National Hardwood Lumber Association which measures the percentage of usable material in each piece. There are five basic hardwood grades as follows: First and Seconds (FAS); Selects; No. 1 Common; No. 2 Common; and No. 3 Common. The Architectural Woodwork Institute (AWI) in their "Quality

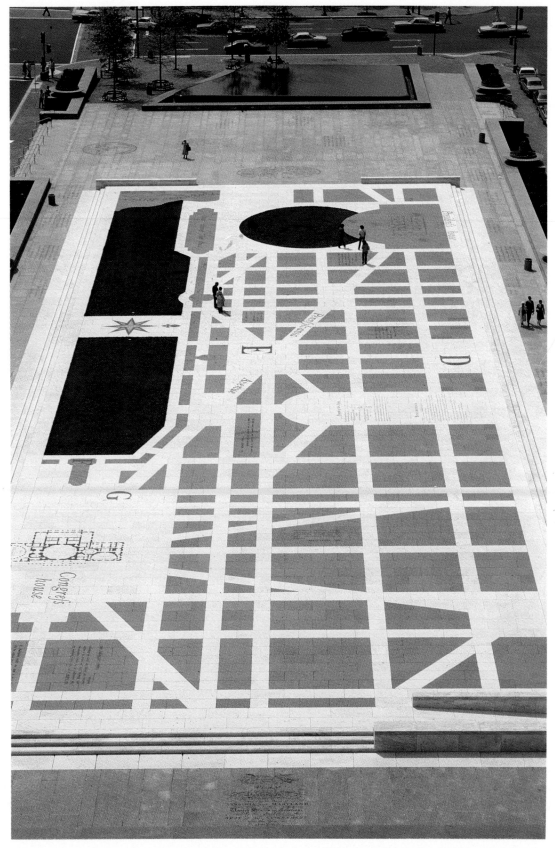

FIGURE 2-1 Pennsylvania Avenue Development, Washington, DC: granite—Sunset Red, Academy, Carnelian, and Bright Red; thermal finish. (George E. Patton/Venturi & Rausch, Architects.) (Courtesy of Cold Spring Granite Company.)

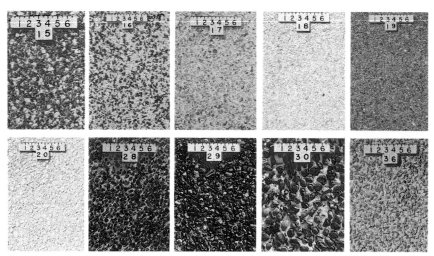

FIGURE 3-3 Panels of various exposed aggregates.
(Used with permission of Portland Cement Association.)

Panel No. 15. Use of three aggregates—verde antique, white marble, and pink granite, on a white background produces a colorful surface.

Panel No. 16. The combination of black obsidian and milky quartz results in a striking contrast.

Panel No. 17. Rose quartz presents a pleasing variety of pink shades against a white matrix.

Panel No. 18. The green of the exposed cordierite aggregate produces a lighter hued surface when viewed from a distance.

Panel No. 19. Pink feldspar is a platy rock that orients itself in the horizontal position during consolidation. The resulting exposed-aggregate surface is very dense and uniform. The large flat crystals reflect light from myriad surfaces.

Panel No. 20. The exposure of very white silica stone on an integrally colored light blue background is most interesting. The distant viewer sees a moderate blue surface with the exposed aggregate barely visible. A closer view shows the mild contrast caused by exposure of the aggregate.

Panel No. 28. Reddish brown Eau Claire gravel contrasts with uniform white background.

Panel No. 29. Use of expanded slag fines results in a coarse-textured white background for the reddish brown Eau Claire aggregate.

Panel No. 30. A chemical surface retarder produced the correct degree of exposure for the ¾-in. (19-mm) to 1-in. (25-mm) Eau Claire aggregate.

Panel No. 36. Exposure of pink marble on white background provides a subtle coloration.

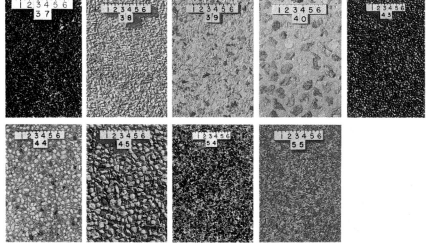

FIGURE 3-3 (continued)

Panel No. 37. Black labradorite rock fractures along crystal faces, and exposed portions sparkle in the sunlight. Aggregate of the size shown here results in a very black surface.

Panel No. 38. White marble in a matrix of silica sand and white cement produces an all-white panel.

Panel No. 39. The brown-yellow-white quartz fractures into flat pieces that orient themselves against the bottom surface of the form. The result is excellent exposure of the quartz.

Panel No. 40. The form was covered with a surface retarder and pink marble was hand-placed uniformly on it. White cement-sand mortar was placed over the marble and consolidated. Backup concrete was then placed and consolidated.

Panel No. 43. Dark expanded shale fines produce a dark background that contrasts with the buff-colored Elgin pea gravel.

Panels No. 43, 44, and 45 demonstrate the effect of increasing the size of coarse aggregate. A range of surface retarders produced the correct reveal for the aggregates.

Panel No. 54. A single-colored green glass was used for the exposed aggregate, yet a two-color effect was obtained due to the difference in size of the translucent aggregates. A special glass that was nonreactive with portland cement was used.

Panel No. 55. Combination of a reflective orange glass with a nonreflective brown natural aggregate makes a very interesting panel.

FIGURE 3-4 Hirshhorn Museum, Washington, DC: Sand-blasted Swenson Pink granite aggregate. Skidmore, Owings & Merrill, Architects.

FIGURE 3-15 Typical precast structure and an insert showing light aggregate exposure using a retarder. (Courtesy of Prestressed Concrete Institute.)

FIGURE 4-10 RBM panel, 39 feet long, being hoisted into place. (Courtesy Brick Institute of America.)

FIGURE 4-23 Wyndham Hotel, Dallas TX: Granite—Sunset Red, polished finish. Dahl/Braden & Chapman, Architects. (Courtesy of Cold Spring Granite Company.)

FIGURE 4-24   National Gallery of Art, East Building, Washington, DC: Tennessee marble. I.M. Pei & Partners, Architects.

FIGURE 4-26   Interior view of Beinecke Rare Book & Manuscript Library, Yale University showing translucent characteristics of white marble. Skidmore Owings & Merrill, Architects. (Photographer, Ezra Stoller © ESTO.)

FIGURE 4-27 LBJ Library, Austin, TX: Travertine facade. Skidmore Owings & Merrill, Architects. (Photographer, Ezra Stoller © ESTO.)

FIGURE 5-5 Ruan Center, Des Moines, IA: example of an unpainted exterior plate wall using weathering steel. Kendall, Griffith, Russel, Artiaga, Architects-Engineers. (Courtesy of U.S. Steel Corp.)

FIGURE 5-6 U.S. Steel Building, Pittsburgh, PA: example of exposed weathering steel facade. Harrison & Abramovitz & Abbe, Architects. (Courtesy of U.S. Steel Corp.)

FIGURE 6-2 Mahogany reception desk. Note the warm, red hues. W.C. Morgan & Associates, Architects. (Used by permission of the Architectural Woodwork Institute.)

FIGURE 6-4 Burl, Carpathian elm.

FIGURE 8-2 Lever House, New York, NY: Blue-green tinted glass used in the forerunner of curtain wall systems. Skidmore, Owings & Merrill, Architects

FIGURE 8-15 First Wisconsin Plaza, Madison, WI: typical sloped glazing. Skidmore Owings & Merrill, Architects.

Standards" grades hardwood species as to the clear areas required and the number of defects allowed. AWI grades are premium, custom, and economy.

FIRE-RETARDANT TREATMENT

The model building codes and most state and municipal codes restrict the use of combustible materials in places of occupancy such as assembly, educational, mercantile, business, institutional, and residential. Since plywood and lumber used for architectural woodwork can be pressure treated with fire-retardant chemicals making them "noncombustible," they are acceptable under most codes for these applications.

Fire-Retardant Chemicals

The salts used for impregnating lumber and plywood through pressure treatment so that they may later receive a transparent finish include ammonium phosphate, ammonium sulphate, borax, boric acid and zinc-chloride.

Since these salts are hygroscopic they cannot be used in locations where the relative humidity exceeds 80%. In addition, some species of lumber such as white oak tend to darken when impregnated with these salts. The pressure treatment plant should be consulted before wood selections are made to ascertain the effect of the fire-retardant chemicals on the specific wood species when a clear finish is required.

Since these chemicals are water soluble, redrying by kilns is necessary after treatment to obtain the required moisture content. The retention of these fire-retardant chemicals in the wood makes sawing, drilling, and woodworking extremely difficult.

Exclusions from Code Requirements

A study commissioned by the Architectural Woodwork Institute (AWI) to analyze the five major model codes with respect to fire resistance requirements for interior architectural woodwork disclosed that certain items were exempt from these requirements.

The five model codes analyzed were as follows:

BOCA	Basic Building Code—1978
ICBO	Uniform Building Code—1979
SBCCI	Standard Building Code—1979
AIA	National Building Code—1976
NFPA 101	Life Safety Code—1976

The exclusions of architectural woodwork from the fire resistance requirements under these codes are the following:

Cabinetry and Casework: These items are considered as furniture or furnishings whether free-standing or fixed or attached and therefore do not require fire resistance treatment.

Wood 1/28 inch or Less in Thickness: Interior finish such as wall and ceiling covering 1/28 inch or less in thickness do not require fire resistance treatment. Paneling that is composed of plywood or particle board with a wood veneer facing can meet this fire resistance requirement if the core material is impregnated with a fire-resistant chemical and an untreated wood veneer, 1/28 inch or less in thickness, is applied to the surface.

Flame Spread Ratings

The model codes and most state and municipal codes refer to ASTM E84 as the test method to determine flame spread requirements (see Chapter 1) for architectural woodwork. There is no unanimity among the codes as to the flame spread ratings for exits, corridors, and other spaces in the various occupancies hereinbefore noted. Each model code or state and municipal code must be reviewed for its requirements.

In addition some insurance companies may require more restrictive flame spread ratings than are required by the model codes or state and municipal codes. ASTM E84 procedures require a 10-minute flame spread test. Factory Mutual and other insurance organizations require that fire-resistant lumber meet American Wood-Preservers' Association (AWPA) Standard C20 and that plywood meet AWPA Standard C27. These standards require that lumber and plywood be subjected to the ASTM E84 test for 30 minutes in lieu of 10 minutes; that the flame spread be not over 25; and that each piece treated be identified with a label by an approved agency such as UL.

STRUCTURAL GLUED LAMINATED TIMBER (GLULAM)

DEFINITION

The term "structural glued laminated timber" (glulam), as noted in U.S. Product Standard PS-56, refers to an engineered, stress-rated product of a timber laminating plant comprising assemblies of specially selected and prepared wood laminations securely bonded together with adhesives. The grain of all laminations is approximately parallel longitudinally. The separate laminations must not exceed 2 inches net thickness. The laminations may be comprised of pieces that are joined to form any length; placed or glued edge-to-edge to make wider ones; or bent to curved form during gluing.

LUMBER

Lumber used for glulam is structurally graded in accordance with standard grading provisions of the species and supplementary requirements of laminating specifications. Some species of lumber used include douglas fir, southern pine, hemlock and larch.

FIRE SAFETY

An advantage of glulam is that the self-insulating qualities of heavy timber sizes cause the assembly to be slow-burning. By proper detailing, fire safety is in-creased, as in the elimination of concealed spaces and the use of fire stops to interfere with the passage of flames up or across a structure.

Building codes generally exempt heavy timber framing from interior finish flame spread requirements, and numerous transparent finishes are available to enhance the color and texture of the wood species selected for glulam construction.

Where fire-retardant chemicals are contemplated, the following factors should be investigated: strength reduction, compatibility of treatment and adhesives, gluing procedures, and fabricating procedures.

WOOD PRESERVATION

Where decay due to fungi, insects, borers, or exposure to high humidity is expected, consideration must be given to pressure-preservative treatment. For the treatment to be effective the proper chemicals must be used and the degree of retention and penetration known. For members exposed to weather and not protected by overhanging roofs or eaves, treatment must be provided or woods resistant to decay must be used.

Pressure-preservative treatment must be used within buildings subject to high humidity levels where the glulam may reach a moisture content in excess of 20%.

Information on pressure-preservative treatment is available in the American Institute of Timber Construction publication AITC 109.

APPEARANCE GRADES

There are three appearance grades for glulam—Industrial, Architectural and Premium. These grades apply to the surfaces of the glulam members and include such items as growth characteristics, inserts, wood fillers and surfacing operations.

Industrial appearance grade is intended for use in industrial plants, warehouses, garages and other uses where appearance is not of primary concern.

Architectural appearance grade is suitable for construction where appearance is an important factor.

Premium appearance grade is used for structures that demand the finest appearance.

Detailed information on appearance grade specifications is contained in AITC 110-83, Standard Appearance Grades for Structural Glued Laminated Timber.

SHAPES

Glulam design and construction make possible the use of structural timbers in a wide variety of sizes and shapes and permit the architect to exercise a wide latitude in the creation of architectural forms which express the structure's function and use.

Glulam timber members may be straight or curved to meet design requirements and specifications and include shapes shown in Figure 6-1.

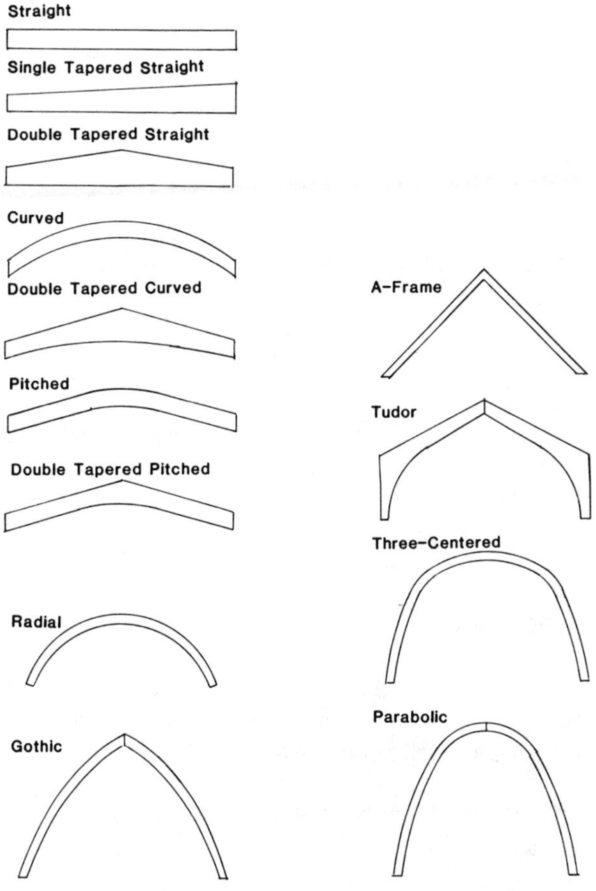

Straight

Single Tapered Straight

Double Tapered Straight

Curved

Double Tapered Curved

Pitched

Double Tapered Pitched

Radial

Gothic

A-Frame

Tudor

Three-Centered

Parabolic

FIGURE 6-1 Structural glued laminated shapes.

DESIGN DATA

Design values, data, and standards relative to the design of glulam are contained in the following publications:

Adhesives	U.S. Dept. of Commerce Product Standard PS 56.
Appearance Grades	AITC 110-83 Standard Appearance Grades for Structural Glued Laminated Timber.
Design Values Softwood	AITC 117, Standard Specifications for Glulam Softwood Species.
Hardwood	AITC 119, Standard Specificiations for Glulam Hardwood Species.
Testing and Inspection	AITC 200 Inspection Manual.
Timber Construction	Timber Construction Manual, published by John Wiley & Sons, Inc.
Wood-Preservative Treatment	AITC 109, Treating Standard for Glulam.

DECKING SYSTEMS

Timber decking is comprised of both laminated and solid timber members. Either system may be employed with glulam beams or arches to complement an architectural expression. Decking is available in a variety of textural exposed surfaces including smooth, saw textured, grooved, straited, and wire-brushed. These may be prefinished in a variety of stains.

Glue laminated decking is assembled from three or more individual kiln-dried laminations into single decking members with T & G patterns. It is available in several thicknesses and appearance grades.

Solid decking is T & G, kiln-dried timber available in select quality grade and commercial quality grade. Wood species include douglas fir, larch, southern pine, hem-fir, and western cedar. The AITC has a publication No. 112 for "Tongue and Groove Heavy Timber Roof Decking" which provides detailed information on solid decking.

HARDWOOD VENEERS

GENERAL

The architect wishing to utilize wood in a design has an infinite variety of wood species from which to select. This is especially true when the architect chooses the veneers of hardwoods which have endless varieties of grain and figure patterns, colors, and textural interest. Some woods have ever-changing highlights and shadows, and others are strong, sturdy, and rugged in character.

FIGURE 6-2 Mahogany reception desk. Note the warm, red hues. W.C. Morgan & Associates, Architects. (Used by permission of the Architectural Woodwork Institute.) This illustration also appears in the color section.

Even within a species, the architect can select from a wide range of figure and grain patterns. Like fingerprints, the grain patterns of any two trees are never exactly alike. This individualism is perhaps one of the most appealing factors when one selects a natural material such as wood as opposed to a synthetic material. Further refinement is achieved by the manner in which the wood is cut, finished and matched. See Figure 6-2.

HARDWOOD SPECIES

Hardwood can be utilized both as a veneer or in the solid form, although for some species the amount of wood is so limited that solid members cannot be obtained in the size and quantity required.

More than 200 species are available from more than 99,000 varieties of hardwoods for use in architectural designs. Hardwood trees, their botanical names, place of origin, characteristics, and uses are shown in Table 6-2.

VENEER TERMS

Veneers from Different Parts of Trees

Veneers are generally cut from the trunk, but some species produce choice and unusual figures from other portions of the growth. See Figure 6-3. Commercially, the various portions of a tree which produce veneers are as follows:

Longwood

Longwood or trunkwood is cut from the trunk and may be rotary cut, half-round rotary, back-cut rotary, flat or quarter sliced or sawn.

Stumpwood

Stumpwood or butt is produced from the stump of a tree. The figure is produced by the wood fibers which,

TABLE 6-2
HARDWOOD (DECIDUOUS) TREES

Species	Location	Color	Use	Pattern
Alder, red (*Alnus rubra*)	West coast of U.S.	Pale pink, brown	Furniture, cabinet work, paneling	Not distinct
Ash, black (*Fraxinius nigra*)	Lake States	Warm brown	Furniture, cabinet work	Clusters of eyes
Ash, White (*Fraxinus americana*)	Eastern U.S.	Cream to very light brown	Furniture	Moderately open grain
Avodire (*Turraenthus africana*)	Liberia, Cameroons	From white to light golden cream	Paneling, furniture, cabinets	Largely figured with a mottle
Birch, sweet (*Betula lenta*)	Adirondacks	Brown, tinged with red	Paneling, cabinetwork, furniture	Grain distinct
Birch, yellow (*Betula alleghaniensis*)	Eastern U.S.	Cream or light brown	Paneling, doors, cabinets, furniture	Plain mild pattern
Butternut (*Juglans cinerea*)	Appalachians	Pale brown	Veneers	Soft to medium texture
Cherry, black (*Prunus serotina*)	Eastern and Central U.S.	Light reddish brown	Furniture, cabinets, veneer panels	Straight grained
Elm, American (*Ulmus americana*)	Eastern and Central U.S.	Light brownish	Furniture, cabinets	Mild grain
Elm, Carpathian Burl (*English Elm*) (*Ulmus campestris*)	France, England, Carpathian Mts.	Brick-red to light tan	Fine cabinetry	Burl
Gum, red (*Liquidamber styraciflua*)	Eastern and Central U.S.	Reddish brown	Cabinetry	Figured
Lauan, red (*Philippine mahogany*) (*Shorea negrosensis*)	Philippines	Red to brown	Furniture, doors, cabinets	Ribbon stripe
Mahogany, African (*Khaya ivorensis*)	Africa	Light pink to reddish brown	Paneling, furniture, cabinets	Figured
Mahogany Tropical American (*Swietenia macrophylla*)	Central and South America	Light reddish to a rich, dark red	Paneling, furniture, cabinets, doors	Figured
Maple, Hard (*Acer saccharum*)	Lake States, Appalachia	Cream to light reddish brown	Furniture, paneling, cabinets	Usually straight, birds eye
Maple, Soft (*Acer saccharinum*)	Lake States, Appalachia	White to reddish brown	Furniture, paneling, cabinets	Straight

(Continued)

TABLE 6-2 (continued)
HARDWOOD (DECIDUOUS) TREES

Species	Location	Color	Use	Pattern
Oak, English Brown *(Quercus robur)*	England	Light tan to deep brown	Paneling, furniture	Noticeable figure and grain
Oak, Red *(Quercus borealis)*	Eastern U.S.	Reddish tinge	Paneling, furniture, cabinets	Flake figure, open pores
Oak, White *(Quercus phellos)*	Western U.S.	Light brown, grayish tinge	paneling, furniture, cabinet	More pronounced than red
Poplar, Yellow *(Liriodendron tulipifera)*	Eastern U.S.	Canary yellow	Paneling, furniture	Straight grain
Rosewood, Brazilian *(Dalbergia nigra)*	Brazil	Dark brown to violet	Paneling, cabinets	Coarse, even grain
Satinwood, Ceylon *(Chloroxylon swietenia)*	Ceylon, India	Pale gold	Paneling, furniture	Rippled, straight stripe
Teak *(Tectona grandis)*	Burma, Java	Tawny yellow to dark brown	Paneling, furniture	Mottled fiddleback
Tigerwood *(Lovoa klaineana)*	West Africa	Gray-brown to gold	Paneling, furniture	Ribbon stripe
Walnut, American Black *(Juglans nigra)*	Midwestern U.S.	Light gray-brown to dark purplish brown	Paneling, furniture, cabinet	Plain to highly figured
Walnut, Circassian *(Juglans regia)*	Europe	Tawny colored	Paneling, furniture	Variegated, streak of black or brown
Zebrawood *(Brachystegia)*	West Africa	Straw and dark brown	Paneling	Striped

compressed in growth, tend to wrinkle as they twist and fold over each other at the point where the roots join the trunk. Stumpwood is generally half-round rotary cut.

Crotchwood

Crotchwood veneers are obtained from the portion of the tree just below the point where it forks into two limbs. Here the grain is crushed and twisted, creating a great variety of plume and flame figures, often resembling a well formed feather. Crotches are sliced parallel to the trunk line, and the outside of the block produces a swirl figure that changes to full crotch figure as the cutting approaches the center of the block.

Burl

Burls come from a warty growth generally caused by some injury to the growing layer just under the bark. This injury causes the growing cells to divide abnormally, creating excess wood that creates little humps. Succeeding growth follows these contours. Cutting across these humps by the half-round method brings them out as little swirl knots or eyes. See Figure 6-4.

Types of Veneer Cuts

Two logs of the same species but with their veneers cut differently will have entirely different visual char-

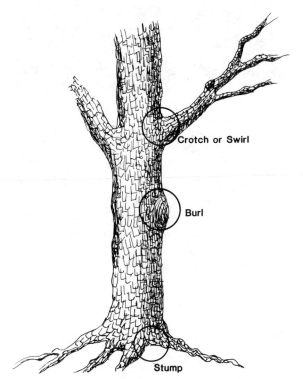

FIGURE 6-3 Veneers from different parts of trees.

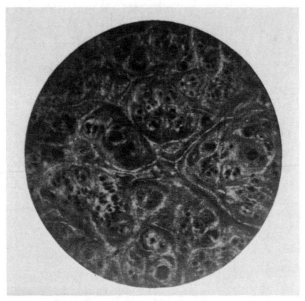

FIGURE 6-4 Burl, Carpathian elm. This illustration also appears in the color section.

acteristics, even though their color values are similar. Six principal methods of veneer cuts are used as follows:

1. *Rotary.* The log is mounted centrally in a lathe and turned against a stationary razor-sharp blade, in a manner similar to the unwinding of a roll of paper. Since this cut follows the log's annular growth rings, a bold, variegated grain marking is produced. Rotary-cut veneer is exceptionally wide, and cannot be sequenced matched.

2. *Half-Round.* This is a variation of rotary cutting in which segments of the log are mounted off-center in the lathe. It results in a cut slightly across the annular growth rings, and shows modified characteristics of both rotary and plain-sliced veneers.

3. *Back-Cut.* The log is mounted as in half-round cutting, except that the bark side faces in toward the lathe center. The veneers so cut are characterized by an enhanced striped figure and sapwood along the edges.

4. *Plain or Flat-Slicing.* The half log is mounted in a movable frame with a stationary knife. The heart-side of the log is placed flat against the guide plate and the slicing is done parallel to a line through the center of the log, producing a variegated figure.

5. *Quarter-Slicing.* The quarter log is mounted as in plain slicing, except that the growth rings of the log strike the knife at approximately right angles, producing a series of straight stripes in some woods and varied stripes in others.

6. *Rift Cut.* This is generally produced from oak species that have medullary ray cells which radiate

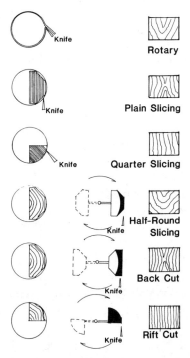

FIGURE 6-5 Methods of veneer cutting.

(A) fiddleback   (B) ribbon stripe

FIGURE 6-6   Figure characteristics of some veneers.

from the center of the log like the curved spokes of a wheel. The rift or comb grain effect is obtained by cutting perpendicularly to these medullary rays either on a lathe or a slicer.

Figure 6-5 shows the six principal methods of cutting.

Figure and Pattern

Figure in veneers is produced by growth rings, flakes or rays, irregular grain pigments, or a combination of two or more of these factors. Typical examples are:

1.   Growth rings—flat cut oak.
2.   Rays—quartered oak or sycamore.
3.   Irregular grain—crotch or stripe mahogany.
4.   Pigment—zebrawood and Macassar ebony.

Figure and pattern terms include the following:

Bird's-Eye: This is due to local sharp depressions in the annual rings, accompanied by considerable fiber distortions. Rotary veneer cuts the depressions crosswise, and shows a series of circlets called bird's eyes.

Blister: This figure is produced by an uneven contour of the annual rings, and not to blisters or pockets in the wood. The veneer must be rotary cut or half-round cut to produce the blistered effect.

Chain: A succession of short cross markings of uniform character remotely suggesting cross-links of a chain.

Cross-Fire: Figures which extend across the grain as mottle, fiddle-back, raindrop and finger-roll are often called cross figure or cross-fire. A pronounced cross-fire adds greatly to the beauty of the veneer.

Curly: Distortion of the fibers produces a wavy or curly effect in the veneer; found mostly in maple or birch.

Fiddle-back: A fine, strong, even, ripply figure as frequently found on the backs of violins, found mostly in mahogany and maple.

(C) cross-fire

(D) mottle

Finger Roll: A wavy pattern in which the waves are about the size of a finger.

Mottle: A variegated pattern which consists principally of irregular, wavy fibers extending for short distances across the face. If a twisted, interwoven grain also has some irregular cross figure, the broken stripe figure becomes a mottle.

Raindrop: When the waves in the fibers occur singly or in groups with considerable intervals between, the figure is called raindrop, as it looks like streaks made by drops striking a window pane at a slant.

Rope: If the twist in the grain of a broken stripe is all in one direction, a rope figure results.

Stripe, Broken: Broken stripe is a modification of ribbon stripe, the markings tapering out and producing a broken ribbon. If the log described in ribbon stripe also has a twist in the grain, the stripes are short or broken.

Stripe, Plain: Alternating lighter and darker stripes, running more or less the length of a flitch and varying in width. It is produced by cutting on the quarter a log that shows growth rings.

Stripe, Ribbon: Wide unbroken stripes, produced by cutting on the quarter, a log from a tropical tree, principally mahogany, with interwoven grain.

See Figure 6-6 for typical examples of figure characteristics of veneers.

VENEER MATCHING

Veneers are matched by utilizing veneers cut from the same log or a segment of a log. The log that produces these veneers is known as a flitch. A flitch is a hewn or sawed log or a section of a log made ready for cutting into veneers. A flitch is also the series of resulting veneer sheets laid together in sequence or as they were sliced or sawn, so that they may be matched and joined to make panel faces.

The joining of veneers in accomplished by means of a "tapeless slicer" which glues the long edges of the veneers together in whatever pattern is to be employed. Certain standard matching patterns are available as follows:

Types of Matching Between Individual Veneer Pieces

Book Match: Every other sheet is turned over, as are the leaves of a book. Thus, the back of one veneer meets the front of the adjacent veneer, producing a matching joint design.

Slip Match: Each sheet is joined side by side and conveys a sense of repeating the flitch figure. Most commonly used in quarter-sliced veneers.

Random Match: A deliberate mismatch by random selection of pieces from one or more flitches, creating a casual or "board like" effect.

End Match: When the panel height exceeds the veneer length, it may be matched vertically as well as horizontally by book matching, thus achieving a uniform grain progression in both directions.

Special Matching Effects: Herringbone, diamond, diamond reverse, V, inverted V.

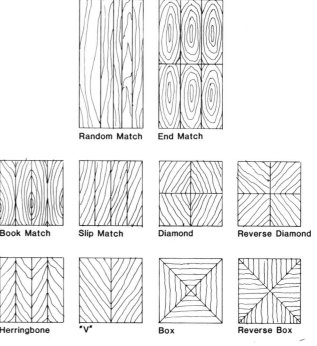

FIGURE 6-7 Matching between individual veneers.

Assembling Veneers Within a Panel

Running Match: Each face is made from as many veneer pieces as necessary. The portion left over is used to start the next panel.

Balance Match: Each face is made with equal width veneers using odd or even pieces. This type of matching may be used in sequence matched panels.

Center Match: Each face is made with an even number of equal-width veneers, which results in a balance of grain and figure on each side of a center line of the panel.

Matching Panels

For architecturally matched panels, either sequence matched panel sets or blueprint matched panels and components should be selected. These types of matching are defined as follows:

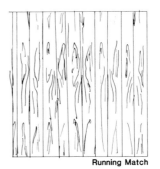

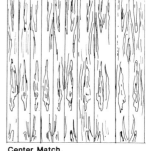

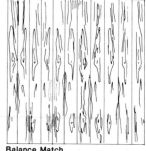

FIGURE 6-8 Assembly of veneers within a panel. (Used by permission of Architectural Woodwork Institute.)

Sequence Matched Panel Sets: These panels are manufactured for a specific installation to a uniform panel width and height. If more than one flitch is required to produce the panels, similar flitches should be used. Doors occurring within the panel run are not sequence matched.

Blueprint Matched Panels and Components: This arrangement produces maximum figure and pattern continuity since all panels, doors, and other veneered components are made to the exact sizes and in exact veneer sequence. The flitch selected should yield sufficient veneers to complete the work.

Figures 6-7 and 6-8 show the various types of veneer matching hereinbefore discussed.

PLYWOOD AND PARTICLE BOARD

PLYWOOD

Construction

Plywood is a general term used to describe a panel made up of an odd number of plies cross-bonded and laminated to one another with adhesive. The outer plies are wood veneers called the face and the back. The grain of these two faces must be parallel to provide stability. This gives the panel nearly equalized strength and minimizes dimensional changes.

The plies immediately adjacent to the outer faces are called cross-bandings. The grain of these plies is usually laid at 90° to the face plies.

Plywood Cores

The cores of plywood can be made of three distinctive types of material as shown in Figure 6-9 and as follows:

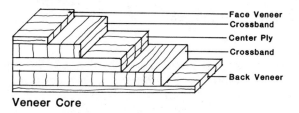
Veneer Core — Face Veneer, Crossband, Center Ply, Crossband, Back Veneer

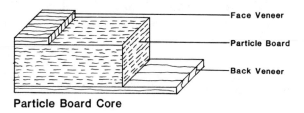
Particle Board Core — Face Veneer, Particle Board, Back Veneer

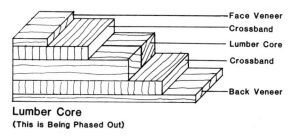
Lumber Core
(This is Being Phased Out) — Face Veneer, Crossband, Lumber Core, Crossband, Back Veneer

FIGURE 6-9 Types of plywood cores. (Used by permission of Architectural Woodwork Institute.)

1. *Veneer Core:* This type of core consists of veneers of wood. Panel thicknesses vary from ⅛ to 1 inch and more and with as little as 3 plies to as many 11 plies or more. This type of core is used primarily for construction and industrial plywood, utilizing softwood lumber and some hardwood species.

2. *Lumber Core:* This core is essentially composed of strips of lumber edge glued into a solid slab using woods of a uniform texture, such as bass and poplar. Lumber core plywood is used primarily for panels, countertops, and panelings. Panels are generally ¾ inch thick but are also available in thicker sizes. The exposed faces are generally hardwood veneers. This core is gradually being phased out.

3. *Particle Board Core:* A core of medium-density particle board (approximately 40 lbs per cu ft) conforming to U.S. Commercial Standard 236, Type 1-B-2 comprises the core for plywood. The face veneers are usually glued directly to the core, although cross-banding may sometimes be used. Particle board core plywood is used for furniture, paneling and countertops.

Plywood Types

Two reference standards issued by the U.S. Department of Commerce govern the various types of plywood used. U.S. Product Standard PS 1-74, Construction and Industrial Plywood is primarily for veneer core plywood embracing over seventy softwoods and about twenty hardwoods. U.S. Product Standard PS 51-71, Hardwood and Decorative Plywood, includes all three cores described previously.

Construction and Industrial Plywood

This category of plywood covers plywood used for the following:

Floors
Subflooring, underlayment, and combination subfloor-underlayment.
Walls
Sheathing, siding, textured interior paneling.
Roofs
Sheathing.
Incidentals
Soffits, cabinets, shelves, built-ins, fences
Concrete Forms
Industrial Applications

Construction and industrial plywood is classified as to exposure capability and grade. Exposure capability is concerned with adhesive durability, and two types are established—interior and exterior plywood—with durability determined by the quality of the adhesive and its resistance to moisture and water. Grade of plywood is a function of veneer grade and panel construction. U.S. Product Standard PS 1-74 covers the wood species, veneer grading, glue bonds, panel construction, dimensions and tolerances, marking, and moisture content.

Specialty plywoods covered under PS 1-74 include marine grade and overlaid plywood. Marine grade is a special exterior type using exterior type glues and has certain restrictions on face patches and core gaps. Overlaid plywood has overlays of fiber impregnated with resin on the face of the plywood panel and includes two types—high-density overlay (HDO) and medium-density overlay (MDO). HDO plywood is suitable for architectural concrete forms. MDO ply-

wood is suitable for painting. Both overlay types are manufactured only in exterior type.

Decorative Hardwood Plywood

This category of plywood comprises the group of fine hardwood architectural plywood that is used for wall paneling, inlays, cabinetry, and furniture.

U.S. Product Standard PS 51-71 establishes requirements for species (three categories based on specific gravity); grades of veneers (by defects); type of plywood (water resistance capability); and construction (type of core).

Veneer grades are defined in PS 51-71 by the permissible defects, which include color variation, mineral streaks, burl, pin knots, and other defects. These are rated as to quality in descending order as follows:

Premium (A)
Good Grade (1)
Sound Grade (2)
Utility Grade (3)
Backing Grade (4)
Specialty Grade (SP)

Type of plywood in PS 51-71 is governed by the resistance of the glue bond to severe service conditions or water resistance capability. The following plywood types are indicative of their water resistance in descending order:

Technical   (Exterior)
Type I   (Exterior)
Type II   (Interior)
Type III   (Interior)

Plywood construction defines the type of core used and PS 51-71 establishes these construction cores as follows:

1. Hardwood veneer
2. Softwood veneer
3. Hardwood lumber
4. Softwood lumber
5. Particle board
6. Hardboard core
7. Special core

Architectural Woodwork Institute (AWI) Standards

AWI standards for interior hardwood plywood are governed by three grades—Premium, Custom, and Economy. The veneer faces for the three AWI grades are governed by the PS 51-71 standard, with AWI Premium Grade requiring PS 51-71 premium (A) faces when exposed two sides and backing grade (4) when only one side is exposed. Refer to AWI Architectural Woodwork Quality Standards for additional information on plywood grades for hardwood panels for Custom and Economy Grades.

Hardwood Panel Sizes

Hardwood panels are normally available in 48-inch widths and 6- to 12-foot lengths in increments of 1 foot. Thicknesses are normally available in dimensions of ⅛, ¼, 5/16, ⅜, ½, ⅝, and ¾ inch. Special sizes up to 5 feet wide and 30 feet long can be fabricated. Availability of sizes in excess of these dimensions should be ascertained from architectural woodwork suppliers prior to detailing and specifying these requirements.

Special Plywood

Marine and overlaid plywood have been described previously under Construction and Industrial Plywoods. Several additional special plywoods are manufactured having certain surface patterns that are of architectural interest. These patterns include the following:

Grooved: An exterior-type fir, cedar, or pine plywood with grooves or kerfs; available factory finished with stains or paint.

Brushed: An exterior or interior plywood produced by subjecting the panels to wire brushing; also available with grooves, and factory finish.

Scratch Sanded: A plywood panel with exposed surfaces scratch sanded to expose texture; available with grooves and factory finish.

PARTICLE BOARD

Particle board is defined as a panel material composed of small, discrete pieces of wood or other lignocellulosic materials that are bonded together in the presence of heat and pressure by a synthetic resin adhesive or other suitable binder. Other materials are added during its manufacture to improve certain properties such as finishing, abrasion resistance, strength, and durability.

Particle board is manufactured in three densities as follows:

Low density: Less than 37 lbs/cu ft.
Medium density: 37 to 50 lbs/cu ft.
High density: Over 50 lbs/cu ft.

Particle board has exceptional dimensional stability, surface flatness and glue-bonding characteristics. Since it can be manufactured from wood chips, shavings, slivers, strands, sawdust, and other by-products of wood, it may ultimately replace veneer core plywood and lumber core plywood as a panel material. Because particle board has excellent dimensional stability, it is being utilized more than ever for hardwood veneer architectural paneling, cabinet work, and countertops.

U.S. Commercial Standard CS 236-66 is a standard for particle board. Type 1-B-2 refers to a medium density core that is normally used for hardwood plywood.

PLASTIC LAMINATE

PRODUCT DESCRIPTION

Manufacture

Plastic laminate is manufactured using a core of phenolic impregnated kraft papers, covered by a melamine resin impregnated pattern sheet and a melamine impregnated translucent overlay sheet. The laminate

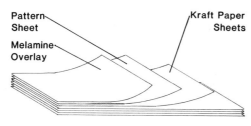

FIGURE 6-10 Plastic laminate veneer

is produced under a pressure of about 1000 lbs/sq in at temperatures in excess of 275° F. The back is sanded to maintain a uniform thickness and to facilitate bonding. See Figure 6-10 for plastic laminate veneer.

Sizes

Plastic laminate is available in widths of 30, 36, 48, and 60 inches and in lengths of 60, 72, 96, 120, and 144 inches. Thickness varies from 0.020 to 0.125 inch.

Finishes and Colors

A variety of finishes is available, including gloss, satin, matte, and a host of textured, embossed, and pebbled finishes. The number of colors is unlimited and includes the solid as well as the striated, marbleized, and wood grain patterns.

Fire-Rated Plastic Laminate

Plastic laminate for use in areas requiring a Class 1 fire-rated finish utilizes a core of fire-retardant, phenolic-impregnated kraft paper. In addition, the backing material to which the fire-rated plastic laminate is adhered must be a fire-rated particle board or hardboard.

STANDARDS

Plastic laminate is manufactured to standards developed by the National Electrical Manufacturer's As-

sociation (NEMA) LD 3-80 and Federal Specification L-P-508H.

The NEMA Standard provides for a number of grades as follows:

> General purpose
> Postforming
> Cabinet liner
> Backer
> Specific purpose
> High wear
> Fire rated

GRADES

The various grades designated in the NEMA Standard have been developed to meet specific requirements based on performance, economy and use. Grades and usages are as follows:

General Purpose Grade: Available in varying thicknesses, GP 50, GP 38, GP 28, and GP 22 (numbers are in mils or thousandths of an inch) with the thinner laminates intended for vertical surfaces such as panels and cabinet doors.

Postforming Grade: PF 42, PF 30, and PF 22 allow simple bends to be made by heating to cover edges and eliminate seams between horizontal and vertical planes.

Cabinet Liner Grade: CL 20 is intended for the interior of cabinets where utility is required and color is less important.

Backer Grade: BK 20 and BK 50 are used for the backs of panels and the undersides of countertops and are intended as a balancing sheet for dimensional stability.

Special Purpose Grade: SP 125 and SP 62 are intended for use where resistance to impact is desirable.

High Wear Grades: HW 120, HW 80, and HW 62 are used for computer access floor tiles where resistance to scuffing, cigarette burns, gouging, and heel marks is essential.

Fire Rated Grades: FR 62, FR 50 and FR 32 are used in areas as previously described under Fire-Rated Plastic Laminates.

PLASTIC LAMINATE ASSEMBLIES

A plastic laminate assembly or panel is very much like plywood, having a laminate face and back, a core, and glue lines to create the assembly.

Adhesives

Several varieties of adhesives are in use to bond laminates to various substrates. These include: (1) thermosetting rigid adhesives such as resorcinol, and epoxy adhesives; (2) contact type adhesives such as neoprene; and (3) hot-melt glues.

Thermosetting adhesives require the use of pressure to insure adherence, and are limited to flat pressable elements. The epoxy adhesives are used primarily for bonding to impervious cores such as steel.

Contact adhesives are suitable for wood and metal surfaces. Both the substrate and the plastic laminate are coated with the adhesive. Care must be exercised in positioning the two surfaces to be bonded, since the components cannot be realigned once contact has been made. Contact adhesives are used on curved surfaces or those not adaptable to press gluing.

Hot-melt glues are primarily for the application of laminate edges using high temperatures and then subjecting the laminate to pressure.

Substrate Materials

The prime requisites for a substrate material to which plastic laminate is adhered are dimensional stability, uniform density, and smooth surface characteristics.

Since plastic laminate is a cellulosic material, it behaves very much like wood and has a grain direction and dimensional characteristics similar to wood. As humidity decreases, plastic laminate tends to contract, and as humidity increases the plastic laminate tends to expand. As a result the following wood substrates offer the best core material for plastic laminate facings: particle board and hardwood faced veneer core plywood.

Particle board, because of its engineered properties, results in a flat core material with the finer chips at the surface, which provides a smooth surface to receive plastic laminate facings. The use of medium-density (40-pound) particle board complying with Commercial Standard CS 236-66, Type 1-B-2 is recommended.

Hardwood-faced veneer core plywood utilizing a Philippine Mahogany face veneer provides a smooth, grain-free surface for laminating purposes. Plywood complying with U.S. 51-71, veneer good grade (1) and backing grade (4) for one surface concealed is recommended.

Fir plywood is not recommended as a core for plastic laminate since the pronounced grain of the fir may result in the telegraphing of the grain in to the plastic laminate. This is particularly true of press-glued laminations.

FABRICATION AND INSTALLATION

The fabrication of plastic laminate components such as panels, cabinets, and countertops should be performed in accordance with NEMA LD-3 requirements and with those of the AWI Architectural Woodwork Quality Standards. Both standards provide quality of work guidelines for gluing, application of edges, cutting, drilling, sawing, and routing.

THERMAL AND MOISTURE PROTECTION

For the most part, the materials utilized for thermal and moisture protection are either concealed (dampproofing, waterproofing, insulation) not generally visible (membrane roofing and fireproofing) or inconspicuous (sealant joints and flashings.) However, some architectural designs and detailing could not be achieved without the use of these materials, which in effect safeguard the structure from the ravages of rain, temperature changes, fire, and water infiltration.

It is because of some of these relatively newer building materials that our more striking architectural concepts are being realized. The curtain wall would not be possible without elastomeric sealants and gaskets. Plaza paving over occupied spaces is more readily waterproofed with the liquid membranes. Upside-down roofing which is less vulnerable to the ravages of weather has been made possible by the development of roof insulations that are not affected by water.

Since these materials make possible unusual designs of modern-day architecture and since some of these materials are relatively new, their characteristics, physical properties, and compatibilities should be examined so that they are better understood when selected and used.

BUILT-UP ROOFING AND WATERPROOFING

A discussion of built-up roofing (BUR) and built-up waterproofing membranes is in order here before the technical aspects are reviewed later. The built-up membranes for both systems are essentially the same, utilizing bitumen-saturated felts and asphalt or coal-tar coatings to produce a layered system that protects a roof or a structure from water penetration. Insulation may or may not be used in conjunction with waterproofing, depending upon the specific requirements. Vapor barriers are sometimes used for BUR. Aggregate surfacing is used only for BUR. The major difference between waterproofing and BUR is that waterproofing is subjected to very little water vapor transmission and is not subjected to the sun, ozone, and temperature changes. As a result, built-up membrane waterproofing, when properly installed, has fewer attendant problems and usually lasts the life of the structure, whereas BUR frequently develops problems and almost always must be maintained and replaced during the life of a structure.

As these systems are examined under the technical discussion it becomes apparent that if BUR were a new material offered to the architect and contractor today, with all its inherent problems, it would receive short shrift from these users. Its use is continued today only because it has been around for a number of centuries and is economical, but its future is less certain as newer roofing systems are being developed.

ROOFING SYSTEMS

Roofing systems are designed to perform several functions such as shedding water, reducing heat and water vapor flow, and providing comfort to occupants and protection to property. Roofs must have the ability to accommodate movement resulting from temperature and structural changes, and the ability to retain their physical properties through all temperature cycles. They should be resistant to traffic, wind, fire, and hail.

The oldest and most widely used roofing system is BUR. In recent years and particularly because of problems associated with BUR and the advent of roofs of unusual geometric contours, new roofing systems have been developed. These include: (1) liquid ap-
plied systems utilizing synthetic rubber or similar elastromers; (2) single-ply roofing using single sheets of polyvinyl chloride (PVC), ethylene propylene diene monomer (EPDM), polyisobutylene (PIB), and neoprene, laid loosely and ballasted or secured by adhesion or mechanical means; and (3) protected membrane roofing or inverted roofing where the roofing is laid on the substrate and insulation and ballast placed over the roofing. While this latter system is called roofing, it is almost akin to waterproofing and may utilize BUR, liquid-applied coatings, or single-ply sheet systems.

BUILT-UP ROOFING

THE SYSTEM

It is important to note that built-up roofing is a system that may consist of the following elements:

1. Vapor barrier
2. Insulation
3. Roofing felts
4. Bitumen
5. Roofing ballast
6. Flashings

It is also important to note that each of the above roofing constituents is produced to materials specifications promulgated by ASTM or other standards. Yet, when a roof failure occurs it is not necessarily a materials failure; most often it is a systems failure. The system fails for a multitude of reasons, which can include the following:

1. Poor roof design
2. Poor quality of work
3. Water vapor transmission
4. Wet substrates and wet felts at time of construction
5. Inadequate flashing design
6. Overheating bitumen
7. Lack of sufficient bitumen between plies

8. Inadequate slopes for drainage

9. Unrestricted movement not compensated by building expansion joints

BUILT-UP ROOFING MATERIALS

Vapor Barriers

See Chapter 1 for a discussion of vapor barriers. The primary function of a vapor barrier in a roofing system is to control the movement of water vapor into the insulation. A vapor barrier or vapor retarder is a material that has a perm rate of not over 1. Most roofing manufacturers require the architect to make decisions on the use of a vapor barrier. As a guideline, the manufacturers suggest that vapor barriers be considered in the following instances: (1) whenever the average January outside temperature is below 45°F; (2) on buildings with excessive moisture conditions such as textile mills, laundries, canneries, breweries, and similar processing plants; and (3) on roofs where the decks contain appreciable amounts of moisture such as lightweight concrete, poured gypsum, asphalt-perlite fills, and similar constructions.

ASTM C755 is a recommended practice for the selection of vapor barriers and outlines the factors to be considered.

For some installations such as cold storage buildings, a vapor stop (a barrier with a perm rating of 0.05) is recommended.

Roofing Bitumens

Bitumen is a generic term describing mixtures of hydrocarbons in viscous or solid form. In the roofing industry it refers to either coal-tar pitch or asphalt.

Overheating bitumen during application of BUR results in the loss of oils by distillation and produces a thin film when applied that does not provide adequate cementing action. The temperature range for heating coal-tar pitch is 325 to 375°F, and for asphalt 400 to 430°F. Underheating below this range results in a loss of application workability and cementing capability.

Coal-Tar Pitch

Coal-tar pitch is produced by the destructive distillation of such materials as coal and wood. Coal-tar pitch has certain cold flow properties which contribute to its ability to self-heal. It is ideal for dead-level or low-incline roofing not exceeding ½ inch in 12 inches. It is also more resistant to water absorption than asphalts. A standard reference for coal-tar pitch is ASTM D450.

Asphalt

Asphalt is found in the natural state but for the most part is obtained from the distillation of petroleum. It is slightly more soluble in water than coal-tar pitch. Asphalts are available having different melt points, with low-melt-point asphalts used on low slopes and higher-melt-point asphalts used on steeper slopes. ASTM D312, a reference standard for roofing asphalt, describes four types, each with a different melt point for varying roof slopes.

Roofing Felts

In a built-up membrane system the roofing felts are used as a reinforcement to stabilize the bitumen layers. They provide the strength required to span irregularities in the substrate and to distribute strains over a greater dimension.

Roofing felts are bitumen saturated and include: (1) organic felts made from wood fiber pulp to which scrap paper and a small percentage of rag have been added; (2) inorganic asbestos fiber felts; and (3) inorganic glass fiber felts and glass mats. The organic felts are saturated with coal-tar pitch or asphalt. The glass mat felts are treated with asphalt or coal-tar. The remaining felts are saturated with asphalt. Table 7-1 shows the various types identified by ASTM designations.

Coated sheets are roofing felts produced from vegetable or animal fibers that have been both saturated and coated on both sides with asphalt. Coated sheets are generally used as a vapor barrier under insulation and as a first ply in a built-up membrane. As a vapor barrier, coated sheets have a perm rating of 0.3. As a first ply in a membrane, coated sheets are intended as protection to the overlying felts above the joints in

TABLE 7-1
ROOFING FELTS

Type	ASTM Designation
Organic felt, asphalt saturated	D226
Organic felt, tar saturated	D227
Asbestos felt, asphalt saturated	D250
Glass fabric, asphalt or tar treated	D1668
Glass mat, asphalt saturated	D2178

rigid board insulation. Coated sheets are produced to meet ASTM D2626.

Roof Insulation

A variety of roof insulations are available and are primarily of the rigid type when used in a BUR system. These include fibrous, granular, and foamed types and various combinations of these types. They are classified by type designation, description, and ASTM number where available, and as shown in Table 7-2.

The various types are also manufactured as composites, usually with polyurethane as one of the constituents because of its excellent insulating characteristics. ASTM C726 is a reference standard for such composite insulation boards.

A more recent innovation in roof deck insulation is the development of tapered insulation. This system ensures that roof inclines will provide positive drainage of water by shaping the insulation rather than designing the structure or adding fill to provide slope. Perlite, cellular glass, polystyrene, and urethane insulations can be utilized for such systems as shown in Figure 7-1.

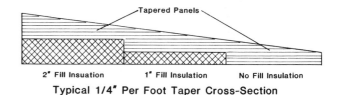

Typical 1/4" Per Foot Taper Cross-Section

FIGURE 7-1   Tapered deck insulation.

Performance of Rigid Insulation

Fiberboard: May absorb moisture and may deteriorate by buckling and warping.

Fiber glass: Will not deteriorate in the presence of water or moisture. Stack vents are effective in permitting water vapor to escape.

Perlite: Is friable and must be handled with care to prevent breakage. It may degrade slightly in the presence of water vapor. Its fire resistance is better than most other insulations.

Polystyrene: Has good resistance to moisture absorption and water vapor. It is combustible and contributes fuel in the event of fire. Its large size reduces the number of joints and the cost of installation.

Polyurethane: Has the best insulating characteristics and is used also as a composite with perlite and fiberglass to achieve the best of these combinations. There is some loss in insulating characteristics with time. Combustibility and toxic fumes resulting from fire demand that safety precautions be exercised.

Cellular glass: Is resistant to the passage of water and water vapor. However, thermal cycling may result in the progressive destruction of the insulation. Cellular glass is strong in compression and is often used where traffic occurs. It has the least insulating characteristics of the insulations available.

TABLE 7-2
INSULATION TYPES

Type	Description	K Factor	ASTM Designation
Fibrous	Wood or cane fiber, felted and compressed with a plastic binder	0.36	C208
Fibrous	Glass fiber, felted and treated with a binder	0.26	—
Granular	Perlite, expanded volcanic rock formed into rigid boards with binder	0.36	C728
Foamed	Polystyrene—Plastic foamed, expanded, and molded into rigid boards	0.24	C578
Foamed	Polyurethane—Plastic foamed, expanded, and molded into rigid boards	0.15	C591
Foamed	Cellular Glass—Ground, remelted, and foamed into rigid boards	0.40	C552

Roofing Aggregate (Ballast)

Roofing aggregate (usually gravel or slag and sometimes marble chips and limestone) is applied to BUR as a surfacing material for a number of reasons. To be effective it must be clean, hard, inert, free of dust, and light in color. It serves to protect the top bitumen coating in the following ways:

1. Protects against destructive action of solar rays.
2. Keeps BUR cooler in sunny weather by reflection.
3. Provides weight against wind uplift.
4. Inhibits weathering of the bitumen and the felts.
5. Reduces fire hazard.
6. Provides limited protection against abrasion from wind.
7. Helps prevent bitumen flow.

Roofing aggregate is specified to meet ASTM D1863. Its gradation is important in order to obtain complete coverage. If too large, gaps will occur in the surfacing; if too small, the surfacing may be easily dislodged and blown off.

PROPERTIES OF BUR SYSTEMS

It has been observed previously that the individual components of BUR are manufactured to standards that have been developed both by ASTM and Federal agencies and yet roofing failures are rife because these are systems failures rather than materials failures. There are probably more studies and literature available on BUR problems than there are on any other building product or system, and the reasons for these failures are attributable to selection, design, quality of work, and job conditions. The variables in the site manufacture of a BUR system are so manifold that only close supervision and inspection will result in a system that can be expected to meet its life expectancy, or perform satisfactorily. Little wonder then that there is a host of panaceas, consultants and substitutes for BUR.

One characteristic in the design of a BUR system is the fact that the components are identified and specified but the system which results from these components is not identified in terms of performance. The development of system criteria for BUR would help overcome some of the shortcomings in the present materials methods of specifying BUR systems. A system-based specification should make reference to specific performance attributes of the overall system such as tensile strength, thermal expansion, thermal movement, flexural strength, tensile fatigue strength, flexural fatigue strength, punching shear strength, impact resistance, and wind and fire resistance. A study outlining these performance criteria and conclusions is contained in National Building Science Series 55, *Preliminary Performance Criteria for Bituminous Membrane Roofing.*

Additional performance criteria for the BUR system are suggested in the NBS study but laboratory test procedures for these procedures were not formulated at the time of the NBS publication. These include notch tensile strength, moisture effects on strength, creep, ply adhesion, abrasion resistance, tear resistance, pliability, permeability, moisture expansion, weather resistance, and fungus attack resistance.

A review of the above performance characteristics is an indication of how earlier building systems were developed and used with very little regard for engineering principles and technical scrutiny. Some semblance of science, must be introduced into the art of building.

PROBLEMS OF BUR SYSTEMS

Water Cutoffs

Over the last fifty years there have been many changes in the application procedures in a BUR system. There was a time when water cutoffs were used to isolate areas of the insulation so that leakage of water into the insulation was contained. Ultimately this practice was discarded and used only as a temporary device to terminate the day's work. Current roofing doctrine emphasizes the importance of permitting trapped moisture to migrate laterally to a point of escape.

Vapor Barriers

Vapor barriers were initially extolled as a virtue and then discarded as a cause for failure. Roofing manufacturers today put the onus for the inclusion of a vapor barrier on the architect. Insulation manufacturers generally regard them as a necessity. Vapor barriers are intended to prevent water vapor from entering into the insulation and then condensing within the insulation and reducing its effectiveness. In addition water vapor may also find its way up between the joints of insulation and cause ridging and blisters in the roof membrane.

Base Sheets

Asphalt coated base sheets were introduced to provide the first ply in a built-up membrane to offset blisters and ridging brought about by the absence of a vapor barrier. However, this violates the principle of mixing coal-tar and asphalt in a coal-tar system since coated base sheets are coated with asphalt. Since the perm ratings of base sheets are low, they are also used as vapor barriers.

Roofing Felts

Saturated felts are used in a built-up system to reinforce the system. They add strength and spanning ability to the membrane but in turn may be subject to membrane failures. Where felts are placed on wet decks (rain, dew, frost, or uncured concrete), they pick up the moisture fairly readily. Moisture changes in the felts will cause swelling and shrinkage. For example, when the bottom felt is not coated with bitumen, as at an insulation joint, moisture or water vapor enters from below and localized dimensional changes due to moisture absorption result in the ridging of the membrane. This ultimately results in a blister. Equally essential is the need to provide complete films of bitumen, to gain good adhesion with no gaps or wrinkles, and to prevent the inclusion of moisture between plies.

Bitumens

The bitumens, especially the asphalts, come from various petroleum sources, and it is difficult to insure uniformity in the grade of asphalt. When installed, the bitumens must be applied uniformly to the surfaces of the felts. Needless to say, the amount applied can vary considerably over a roof deck. Bitumens are also subject to degradation from the sun, temperature changes, UV, and wind erosion.

Other Problems

Additional problems areas for BUR systems include: (1) lack of slope that leads to ponding; (2) placement of mechanical equipment on roofs that require mechanics to traverse the roof for maintenance of the equipment; and (3) installation of roofs during inclement weather; necessitated by construction schedules.

Incompatibility between different bitumens is also important. These incompatibilities are evident by the appearance of oil spots and staining. When coal tar and asphalt are incompatible, one will be softened and the other hardened, the nature and extent of incompatibility depending upon the chemical composition and physical structure of both. Where doubt exists as to the compatibility of the bitumens, a test method, ASTM D1370 may be used to check the incompatibility.

INSULATED METAL DECKS AND BUR

For metal roof decks utilizing vapor barriers, insulation and BUR there are certain recommendations that have been developed by Factory Mutual System that are concerned with the components and methods of installation.

Loss Prevention Data 1-28, published by Factory Mutual System, describes acceptable vapor barriers and adhesives and acceptable insulations and methods of fastening. These recommendations govern Class I and Class II assemblies, with respect to wind uplift and fire hazard.

PROTECTED MEMBRANE ROOFING

A protected membrane roof or inverted roofing is essentially an upside-down roof system where the

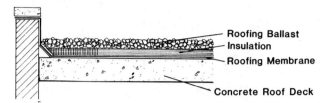

FIGURE 7-2 Inverted insulated protected roof systems.

waterproofing membrane is installed directly on the substrate, with the insulation and ballast applied above the membrane. This system is simply a rearrangement of the constituent elements of a roofing system to alleviate the problems encountered by BUR when it is exposed to the weather. While we call the system a protected membrane roofing system, it is in reality nothing more than what is normally referred to as membrane waterproofing. See Figure 7-2.

There is no need for a separate vapor barrier, since the membrane acts additionally as a vapor barrier by virtue of its location.

Because of its configuration with the membrane concealed and protected thermally, the waterproofing membrane becomes less vulnerable to a variety of stresses. The stresses which contribute to aging or sudden damage are eliminated. These stresses include heat aging, rapid temperature changes, sunlight, and exposure. Hail, fire, and wind erosion are no longer to be endured with protected membranes. Foot traffic to maintain roof equipment cannot injure the membrane, since it is no longer exposed.

INSULATION REQUIREMENTS

The introduction of protected membrane roofing could not occur until a suitable insulation which would be impervious to water and which would retain at least 80% of its insulating characteristics was developed. With the advent of foamed polystyrene as an insulation material, the protected membrane roof system became viable.

New detailing is required to permit protected membrane roof systems to drain properly. A slope is essential at the membrane surface so that the insulation does not lie submerged in water, thereby affecting its insulating characteristics. Slightly open joints between insulation boards, chamfered bottom edges, and grooved bottoms permit water to flow more easily to drains. By designing for about 15% more insulation the effects of wet insulation on thermal transmission is overcome.

MEMBRANE FOR PROTECTED MEMBRANE ROOFING

Initially, with the advent of protected membrane roof systems, the membrane most widely used was the built-up bituminous asphalt roofing membrane. Over the years, as liquid applied waterproofing systems and single-ply sheet systems were developed, they have been utilized as waterproofing membranes for protected membrane roofing. See liquid-applied waterproofing systems and single-ply sheet roofing systems described hereinafter.

SINGLE-PLY SHEET ROOFING

TYPES AND APPLICATIONS

Single-Ply Types

As noted hereinbefore, the problems associated with built-up roofing have given rise to the development of synthetic sheet single ply roofing systems. As with the development of sealants during the late 1950s and 1960s, which required practical application techniques and product standardization to effectively stabilize sealant technology, so too are we witnessing today a proliferation of single-sheet polymeric roofing materials of various chemical formulations and application techniques. In due time the manufacturing industry together with architects and applicators will sort out the poor performers and concentrate on those materials and applications that offer more reliable performance. All of the single-ply sheets discussed herein are also capable of being used for inverted roofing systems.

The current materials that comprise the more promising candidates are the following:

Neoprene
Ethylene-propylene-diene-monomer (EPDM)
Chlorosulfonated polyethylene (CSPE)
Chlorinated polyethylene (CPE)
Polyisobutylene (PIB)
Polyvinylchloride (PVC)
Laminates and composites (modified bitumens)

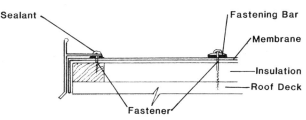

FIGURE 7-4   Mechanically attached single-ply roof system.

Application Techniques

At present, in addition to the various types of single-sheet formulations, there are a number of application methods that are utilized and recommended by the single-ply roofing manufacturers. Not all of the single-ply sheets are installed using all of the techniques described below. Manufacturers' products and recommended installation techniques must be investigated and followed in order to avail oneself of the warranties offered. The installation techniques are as follows:

Loose-Laid and Ballasted

This system provides for the application of the sheet membranes laid loosely and with the following vulnerable conditions secured either mechanically or by adhesions: (1) perimeters; (2) roof projections; and (3) expansion joints. The roof is then ballasted with smooth washed river bed stone or with pavers. Stone ballast may not be permitted by code in hurricane areas. See Figure 7-3 for typical loose-laid system.

Mechanical Attachment

With this system, single-ply membranes are secured at intervals to the roof deck system through metal disks, or by metal or plastic bars placed at lap joints and secured to the roof deck. See Figure 7-4 for typical mechanically attached systems.

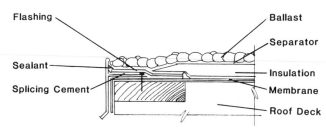

FIGURE 7-3   Loose-laid and ballasted single-ply system.

Bonded or Adhered

In this system the single-ply membrane is bonded or adhered partially or fully to the substrate with bitumen or a compatible special adhesive. This system is used where the additional weight of ballast cannot be used and where the geometry of the roof is such that it cannot maintain ballast.

In some situations combinations of these systems are recommended where both mechanical attachment and spot adhesive is used. Various arguments are made relative to the numerous attachment systems. There are those who desire in the loose-laid system to permit uninhibited expansion and contraction of the single-ply sheet. There are those who extoll the virtues of complete bonding to detect and correct leaks, since in a bonded system lateral water movement is improbable.

Lap Joints

Joints in single-ply roofing are made by heat, fusion, torching, solvent welding, tacky tapes, or adhesives.

PROPERTIES FOR SINGLE-PLY ROOFING

As always with the development of new systems, it takes time for standards-making bodies such as ASTM, ANSI, and governmental agencies to develop and promulgate standards that would govern both the materials requirements and the installation requirements of single-ply roofing systems.

The architect and contractor must make choices today on the basis of manufacturers' claims and histories of installations inasmuch as standards for materials and installations have not yet been standardized. Pertinent to selection of a specific material would be an investigation of the following physical properties:

Tensile strength
UV resistance
Low-temperature brittleness
Ozone resistance
Elongation
Tear resistance
Water vapor transmission
Weathering resistance
Abrasion resistance
Puncture resistance
Heat aging
Fire resistance

Several of these properties may be investigated utilizing ASTM D3105.

Concurrent with the analysis of the properties of materials would be the field application techniques such as procedures for making lap joints. Does the procedure require intricate, sophisticated use of welds to be performed by specially trained personnel, or can it be made with commonplace adhesives by roofers? Some laps can be made with adhesives; others require heat fusion, torching, solvent welding, or tacky tapes. Formulators and users of these synthetic rubbers and elastomers must not only take into account the properties of the materials but must also be conscious of the field application problems and the compatibility of the materials with other materials of the roofing system.

SINGLE-PLY SHEETS

Sizes

Sheet membranes are generally available in roll form in various lengths, and in widths of 36 inches minimum and thicknesses of 30 to 60 mil. For some installations, manufacturers will prefabricate or seam sheets to as much as 45 feet wide and 150 feet long.

Neoprene

Neoprene is the oldest of the synthetic rubbers, having been introduced in the early 1930s. Neoprene has

been formulated and used for a variety of purposes. It is based on polymers of chloroprene, and has excellent aging characteristics. Properly formulated, it is resistant to ozone, sun, weathering, UV, and temperature extremes, and will not support combustion. It can be installed using all three systems described hereinbefore. It is at present one of the more expensive synthetic sheets.

Ethylene-Propylene-Diene Monomer (EPDM)

EPDM synthesized from ethylene, propylene, and diene monomer has been available since about 1963, when it was introduced to overcome some of the shortcomings of butyl rubber. It has good UV and weathering characteristics and lap joints are readily made with special adhesives. It is at present one of the more widely used synthetic sheets.

There is a proposed standard being considered for EPDM; some of the properties suggested are shown in Table 7-3.

Chlorosulfonated Polyethylene (CSPE)

CSPE single-ply membranes are produced from DuPont "Hypalon", a synthetic material that has been available since the 1950s when it was applied in liquid form over neoprene to provide ozone resistance and color. The fully vulcanized sheet material retains its

TABLE 7-3
PROPOSED PROPERTIES FOR EPDM

Property	Test	Requirement
Tensile strength	D412	1300 psi, min.
Elongation	D412	300%, min.
Tear resistance	D624, Die C	125 pli, min.
Dimensional stability	D1204	1–2% max.
Brittleness temp	D746	−67°F min.
Ozone resistance	D1149	No cracks
Water absorption	D471	2% max.
Permanent set	D412	20% max.
Permeability	E96, Procedure A	2.0 perm max.

flexibility and toughness over a long period of time. Seams are made by heat sealing without the use of adhesives or cements. The sheets are reinforced with polyester scrim. Hypalon is basically a chlorinated polyethylene containing chlorosulfonyl groups with a high molecular weight, and low-density polyethylene.

Chlorinated Polyethylene (CPE)

CPE synthetic sheets are available with or without polyester reinforcement. These synthetics are resistant to hydrocarbons and air-borne pollutants. The materials have been exposed to extremes of weather in Arizona and Florida for over ten years. These sheets are composed of high molecular weight, low-density polyethylene that has been chlorinated.

Polyisobutylene (PIB)

PIB is a synthetic sheet material generally bonded to a nonwoven synthetic felt reinforcement that permits installation using asphalts or contact adhesive. It has excellent weathering characteristics.

Polyvinylchloride (PVC)

PVC is made of PVC resin and plasticizers and is available with or without reinforcement. It has been in use for over 20 years. Some nonreinforced PVC will shrink as much as 5%. Weathering ability is less than that of Neoprene and EPDM. It is not compatible with tar or asphalt. Lap seams are made by solvent welding, and sealing.

Laminates and Composites (Modified Bitumens)

In addition to the synthetic roofing membranes described above, produced by a number of manufacturers, there are proprietary single-ply membranes that are made by the addition of *rubberizing* ingredients to asphaltic materials which are similar to those used in ordinary built-up roofing.

In the manufacture of modified bitumens, a synthetic polymer such as polypropylene or styrene-butadiene is blended with asphalt. The manufacturing process is critical, since good dispersion of the rubber and the bitumen is essential to obtain a good quality, single-ply membrane. Some products of this marriage produce a tacky membrane which allows the material to be self-adhering. Others produce combination laminates using polyethylene sheets, reinforcing materials, and aluminum foil in the composites.

The advantage of these products is that they are similar to built-up roofing and can be installed with asphalt so that roofing contractors using their regular crews can install the product easily with only minor retraining.

FACTORY MUTUAL RECOMMENDATION

For loose-laid systems and adhered, single-ply membrane roof systems, Factory Mutual Engineering Company recommends certain installation requirements when windstorms, combustibility, and leakage must be considered. These provisions are contained in Loss Prevention Data 1-29.

In Zone 3, which includes areas of the country where the wind velocity pressure is 47 psf or above, FM recommends that paving blocks be used over loose-laid systems rather than gravel.

LIQUID-APPLIED ROOFING MEMBRANES

TYPES

In the discussion on built-up roofing, the insulations described were confined to the rigid board type. However, with the advent and use of foamed-in-place urethane insulation, a host of liquid-applied roofing materials have been developed. Here too, the proliferation of liquid-applied roofing membranes are varied in their chemical formulations, and only time will dictate which of these will survive as long-term viable and economical systems.

Some of the present liquid-applied roofing membranes are the following:

Acrylic
Butyl
Hypalon (chlorosulfonated polyethylene)
Silicone
Urethanes (aliphatic and aromatic)
Vinyl

The reasons for the application of liquid applied roofing membranes over foamed-in-place urethane stem from the fact that the insulation must be protected from the adverse effects of UV, moisture, water vapor, water, and weathering.

COATING SELECTION

There are few if any standards that the architect or contractor can utilize in making decisions on the selection of a liquid membrane for use over foamed-in-place urethane. There are, however, a number of factors to consider when making a selection. Some of these factors are the geographical location and the local climatological environment, chemical and industrial exposure, physical abuse resistance, ease of maintenance, aesthetics, application conditions, and economics.

Northern cold climates require a coating with low temperature flexibility, where southern warmer climates require a coating possessing a high UV resistance. Rainy areas need a coating with low permeability.

PROPERTIES

Some of the more important physical characteristics to review in making an evaluation of a specific liquid applied roofing material are the following:

Tensile strength
Elongation
Adhesion

Water vapor transmission
Weathering resistance
UV resistance
Tear resistance
Flammability
Heat aging
Impact resistance
Ozone resistance
Abrasion resistance

Only Hypalon, chlorosulfonated polyethylene, has a reference standard that has been developed at this time, ASTM D3468, which may be used to determine physical properties.

In addition, an ASTM standard D3105 may be used for procedures and test methods in the evaluation of liquid-applied systems.

LIQUID-APPLIED ROOFING SYSTEMS

Liquid-applied roofing systems are available in both single-component types and two-component types. They are applied by brush, roller, squeegee, trowel, or spraying to form a seamless waterproofing membrane. The single-component types solidify through evaporation of a solvent or water or by chemical cure. The two-component types solidify by chemical cure. The uniformity of thickness and freedom from bubbles and pinholing obtained with a liquid system is a function of the quality of work involved and the surface roughness of the substrate.

Acrylic

These liquid-applied systems are derived from acrylic or methacrylic esters and have good resistance to UV and weathering. They are one-component systems and are available in a number of colors.

Butyl

This system is formulated by synthesizing isobutylene with isoprene or butadiene. It is resistant to ozone and weathering and has low water vapor permeability. It has good resistance to corrosive chemicals but poor resistance to oils. It is a two-component system.

Hypalon (Chlorosulfonated Polyethylene)

This system has been available for quite some time and is available as a one-component system, in a variety of colors. It has outstanding resistance to ozone, heat, and weathering and has good color retention. It is often used as a colored coating for other elastomeric membranes, notably neoprene.

Silicone

These materials consist of semiorganic polymers which are resistant to high temperatures and are flexible at low temperatures. They provide good resistance to oxidation, ozone, and weathering and have a high water vapor permeability. They are available as both one- and two-component systems. They do have a tendency to retain dirt and may darken with exposure.

Urethane

Urethanes constitute a class of synthetic polymers produced from the reaction of di-isocyanate and a poly-ol. Urethanes may be formulated with widely varying properties, including the ability to be colored. Properly formulated, they have good resistance to weathering, ozone, oils, and solvents. They are available as one- or two-component systems.

Vinyl

The vinyls are available as one-component systems. Through proper formulation they can be compounded to provide resistance to weathering. They are not compatible with asphalt or tar. They are available in light colors.

WATERPROOFING SYSTEMS

The term *waterproofing* is defined in ASTM D1079 as the "treatment of a surface or structure to prevent the passage of water under hydrostatic pressure." Since the presence of unwanted water inside a structure is a visible, annoying, and damaging element, a host of waterproofing materials and systems have been spawned for a market eager to find a panacea against water infiltration.

The types of waterproofing systems available can be classified as shown in Table 7-4.

Most waterproofing systems are designed for application on exterior surfaces, since they form a barrier to the entrance of water. The walls and floors provide the support system to counterbalance the hydrostatic head of water. Conversely, where reservoirs, tanks or pools are waterproofed to contain water, the waterproof linings are installed on the water side of the structures for similar reasons. In these instances the waterproofing is said to be applied to the "positive side" or "positive face."

However, several waterproofing systems are designed to be applied on the side opposite the potential source of water (the "negative side"), since the systems also have the structural ability to withstand the hydrostatic head of water.

BITUMINOUS BUILT-UP MEMBRANES

Materials

The bitumens and felts used for built-up membranes are essentially the same as hereinbefore described for built-up roofing systems, except that the reference standard for asphalt for waterproofing is ASTM D449.

Bitumen-saturated cotton fabric, ASTM D173, or bitumen-saturated glass fabric, ASTM D1668, is recommended for reinforcement of membrane water-

TABLE 7-4
TYPES OF WATERPROOFING SYSTEMS

Type No.	System
1	*Bituminous Built-up Membrane* Asphalt or coal tar.
2	*Single-Ply Sheet Materials* Neoprene Butyl EPDM (Ethylene-propylene-diene-monomer) PVC CPE Rubberized asphalt laminate (modified bitumen)
3	*Liquid-Applied Membranes* *(for application between slabs and below grade walls)* Polyurethane, 1- and 2-part Tar-modified polyurethane Rubberized asphalt.
4	*Surface-Applied Liquid Membrane* *(traffic-bearing slabs)* Polyurethane.
5	*Metallic* Iron filings mixed with cement and water.
6	*Bentonite* A granular mineral that forms a water-repelling gel when in contact with moisture.
7	*Cemetitious (Crystalline) Systems* Mixtures of cements and chemical additives, which are absorbed into concrete by water crystallization which fills concrete pores and capillaries.
8	*Integral Admixtures* Proprietary admixtures added to concrete mixes to render the concrete watertight.
9	*Proprietary Systems* Various coating, sheets, powders, etc., that are claimed by their respective manufacturers to protect against passage of water.
10	*Cementitious Coatings* Cement with proprietary admixtures, which form a film surface for positive or negative surface waterproofing.

TABLE 7-5
PLIES OF MEMBRANE VS. HYDROSTATIC HEAD

Head of Water in feet	Number of Plies	Bitumen Moppings
1–3	2	3
4–10	3	4
11–25	4	5
26–50	5	6
51–100	6	7

proofing at internal and external corners or other points subject to strain and for the flashing of all projections through the membrane.

Do not mix bituminous materials. Use either coal tar systems or asphalt systems throughout.

Number of Plies vs. Hydrostatic Head

In the determination of the number of plies of built-up membrane to be used with respect to the hydrostatic head of water to be encountered, the architect or contractor may be guided by the data shown in Table 7-5.

Weight of Bitumen Coatings

In the application of bitumen the architect or contractor may utilize the amounts shown in Table 7-6 for each 100 square feet of membrane.

SINGLE-PLY SHEET MEMBRANES

The single-ply sheet materials used for waterproofing are identified under Single-ply Sheet Roofing.

TABLE 7-6
BITUMEN PER SQUARE

	Each ply	Final coating
Asphalt	30 pounds	60 pounds
Coal tar	25 pounds	70 pounds

The characteristics and properties of neoprene, EPDM, PVC, and CPE are discussed hereinbefore under Single-ply Sheet Roofing.

Butyl

Butyl rubber is an elastomer synthesized by copolymerizing isobutylene with isoprene or butadiene. It has extremely low water vapor permeability and is resistant to ozone and weathering. Its major flaw is the adhesion of lap seams which require special materials and workmanship to achieve good results.

Rubberized Asphalt Laminate

Rubberized asphalt is a mixture of reclaimed ground rubber and asphalt surfaced on one side with polyethylene so that the unsurfaced rubberized asphalt becomes self-adhering when used as waterproofing.

LIQUID-APPLIED WATERPROOFING MEMBRANE (CONCEALED)

Liquid-applied waterproofing membranes for use between a horizontal wearing slab and a substrate and application on walls below grade are shown in Table 7-4. ASTM Standard C836 sets certain criteria for materials designed for use between horizontal slabs, such as: (1) low-temperature flexibility and crack bridging; (2) adhesion-in-peel after water immersion; and (3) extensibility after heat aging. Another ASTM standard, C898, recommends application procedures for horizontal between-deck use of liquid-applied systems.

The advantages of a liquid-applied system are: (1) there is a continuity of the membrane on horizontal and vertical planes, around projections and geometric angles; and (2) with complete adhesion of membrane material to substrate, leaks are more easily found and isolated since water cannot easily travel beneath the membrane laterally and vertically.

Polyurethanes

The major products in use today as liquid-applied membranes are the polyurethanes—both the one- and two-component types and those modified with coal tar pitch. The characteristics of the polyurethanes are discussed hereinbefore under Liquid-Applied Roofing Systems.

Generally, the polyurethane systems are applied cold to a thickness of approximately 60 mils by means of trowel, squeegee, roller, brush, or spray.

Rubberized Asphalt

Rubberized asphalt is a combination of ground-reclaimed rubber and asphalt. It is usually applied hot to an average thickness of 180 mils.

SURFACE-APPLIED LIQUID MEMBRANES

Surface-applied liquid membranes have been formulated to meet the requirements necessary to waterproof traffic-bearing slabs for either vehicular or pedestrian traffic. This includes such structures as parking garage decks and walkways, where hydrostatic heads of water are not encountered but where by virtue of their exposure the decks require waterproofing.

The materials employed are primarily one- and two-component urethanes with some aggregate broadcast in the surface for traction. ASTM C957 is a materials standard for high solids content, cold liquid-applied elastomeric waterproofing membrane with integral wearing surface.

Some of the criteria set forth in the ASTM standard for those products which may be utilized for this purpose are physical requirements such as (1) low-temperature flexibility and crack bridging; (2) adhesion-peel resistance after water immersion; (3) chemical resistance; and (4) weathering resistance and recovery from elongation.

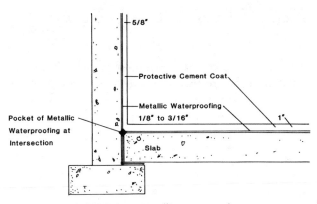

FIGURE 7-5 Metallic waterproofing.

METALLIC WATERPROOFING

Metallic waterproofing is a system of waterproofing the interior surfaces of below-grade concrete walls and floors that may be subject to hydrostatic heads of water. It is usually employed in confined sites or where it is impossible to apply membranes on exterior surfaces, as when buildings adjoin one another.

The system consists of applying brush coats of finely powdered iron mixed with portland cement and water on concrete surfaces. The system is built up to about $\frac{1}{8}$ to $\frac{3}{16}$ inch thick and then covered with a protective coat of mortar. See Figure 7-5 for typical metallic waterproofing detail.

This method of waterproofing is based on the fact that when water enters the system, the powdered iron expands and prevents the intrusion of water. The mechanical bond of the metallic waterproofing to the concrete substrate allows the system to be installed opposite the side from which the hydrostatic water head exists.

BENTONITE WATERPROOFING

Bentonite is a high-swelling type of clay composed principally of the mineral montmorillonite and small amounts of feldspar, mica, and volcanic ash. It has the capability of absorbing a considerable amount of water and swells accordingly, forming a gel. It is this gelling property which permits it to stop water infiltration when it is converted to a gel by the presence of water.

Some manufacturers insert the dry bentonite in the flutes of a biodegradable kraft panel which is affixed to the exterior face of foundation walls or placed below concrete slabs. Others formulate powders which when mixed with water are either trowel applied or sprayed to surfaces to be waterproofed and then backfilled immediately.

The application is such that the bentonite is always placed on the water side of the structure.

CRYSTALLINE CEMENTITIOUS WATERPROOFING

Crystalline cementitious waterproofing is a concept based on the process of crystallization. The proprietary chemicals contained in the formulations react with water and other elements in concrete to form crystals which penetrate the concrete and fill pores and capillaries. The crystals block the passage of water and become an integral part of the concrete.

This system has been used in Europe for some time, primarily in water tunnels, and has been applied to structures as a result of its effectiveness. While intended to be placed on the wet side of a structure, it may be applied to the interior face when existing conditions preclude exterior application, provided it is wetted for some time after application to promote the crystallization.

SELECTING THE WATERPROOFING SYSTEM

In making a decision on the type of waterproofing system to be selected, the designer should understand the nature of the waterproofing problem and its location in the structure and then make an assessment of the systems available to arrive at a viable solution.

A structure may be subject to water infiltration at a number of vulnerable locations. It may be at foundations below grade due to a hydrostatic head of water. It could occur at a plaza with below-grade spaces requiring protection. The structure could be vulnerable due to internal equipment such as pumps

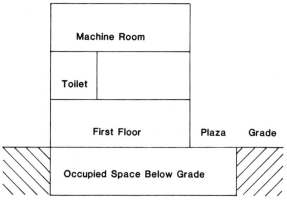

FIGURE 7-6 Areas of building subject to water.

in mechanical spaces or plumbing fixtures in toilet rooms. At each of these locations the type of waterproofing to be selected may be different. Figure 7-6 shows a typical structure where these problems could occur.

Referring to Figure 7-6, at the basement below grade, if the water table were above the basement floor level several waterproofing systems could be applied to the exterior face of the foundation wall if the structure were being constructed in an unconfined area so as to be accessible from the exterior. Reviewing Table 7-4, waterproofing systems Types 1, 2, 3, 6, 7, and 8 are possible systems to be evaluated. In addition, it would be advantageous to learn what systems local waterproofers are using in the particular locale and with what success. Quite often newer and more sophisticated systems are less frequently used in remote areas because of unfamiliarity with the system.

Where the foundation of a new structure is hemmed in because it is in a built-up area, and waterproofing can be applied to the inside face only, Table 7-4 would suggest that only Types 5, 7, and 8 are possible solutions.

At the plaza level in Figure 7-6, waterproofing systems in Table 7-4 would suggest that Types 1, 2, and 3 may be utilized after evaluations are made relative to the ability of waterproofers to handle the specific system selected.

For use at the toilet rooms in Figure 7-6, there are new liquid-applied waterproofing systems that can be installed with thin-set tile to form a waterproof membrane.

At the machine room level in Figure 7-6, a Type 4 (Table 7-4) traffic bearing system applied to the surface would eliminate the need for double-slabbing with concealed waterproofing.

PREPARING THE SURFACE

Where waterproofing systems employing built-up membrane, single-ply membranes, and liquid membranes are intended, the surfaces receiving these waterproofing systems must be prepared so as to present a uniform oil-free surface, free of concrete projections and honeycomb or pits.

Projections of concrete should be ground down and honeycomb and pits filled with mortar so that when these waterproofing systems are applied there will be a solid backing. Failure to observe these requirements may result in puncturing or bursting of the membrane.

PROTECTING THE MEMBRANE

Since the waterproofing membrane may be damaged by other construction trades during the building process, it is good practice to provide some form of protection to both vertical and horizontal waterproofing and to include these provisions in the design.

Typically, for vertical applications a fiberboard type of material is applied to the face of the waterproofing to guard against earth backfilling operations where stones may puncture the membrane unless properly protected.

For horizontal protection an asphaltic, multi-ply, semi rigid board is placed over the membrane to protect it from construction traffic.

The built-up, single-ply, and liquid-applied systems must always receive the above described protection. Metallic waterproofing, as noted previously, is protected with mortar.

SHEET METAL

Information on metals used for ornamental metalwork is contained in Chapter 5, Metals. It includes basic information on aluminum, copper, stainless steel and steel that may be informative to the reader when reviewing the requirements of these metals for applications to roofing and flashing.

SHEET-METAL FABRICATION TECHNIQUES

Seams and Joints

Metals are joined with seams in a number of ways depending upon the metals and their application. A seam or joint in sheet-metal work may be rigid or loose, depending upon its intended function. Rigid seams are developed by: (1) fastening with rivets, screws, or bolts; or (2) soldering; or (3) fastening and soldering in combination.

Loose seams and joints are utilized to permit expansion and contraction of sheet-metal work. They work in one of two ways, either by sliding or flexing. The loose lock seam is an example of the sliding type. The batten seam is an example of the flexing type.

Watertightness in a seam is best obtained and is more positive with soldering, although modern day sealants such as urethanes, silicones, polysulfides and butyls are being used successfully for watertightness where a loose joint is required. Seams should be oriented properly to shed water. See Figure 7-7 for example of typical sheet metal seams, locks, and joints.

Soldering

Soldering in sheet metal work is used to join metals using a filler alloy (solder) which melts and flows below 800°F. Since these temperatures are comparatively low, there is no alloying action between the solder and the base metals being joined.

Of all the metals used for sheet metal work only aluminum cannot be soldered successfully. Mechanical joining is recommended for aluminum where strength is required to be transferred and a sealant used in the joint if watertightness is essential.

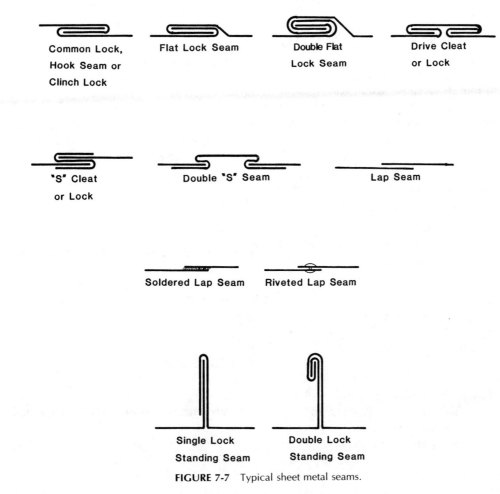

Common Lock, Hook Seam or Clinch Lock

Flat Lock Seam

Double Flat Lock Seam

Drive Cleat or Lock

"S" Cleat or Lock

Double "S" Seam

Lap Seam

Soldered Lap Seam

Riveted Lap Seam

Single Lock Standing Seam

Double Lock Standing Seam

FIGURE 7-7 Typical sheet metal seams.

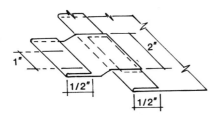

FIGURE 7-8 Typical sheet metal cleat.

Cleating

While sheet metal work is attached to a substrate by either fasteners or cleats, cleating is a preferred method because it permits movement of the sheet metal work without buckling. Cleats are usually of the same metal and thickness as the base metal, are usually 2 inches wide by 3 inches long and are generally spaced 12 inches on center. See Figure 7-8 for typical cleat.

FASTENERS FOR SHEET METAL

Since corrosion due to galvanic action can occur if dissimilar metals are used in sheet metal work, the following guidelines are suggested for fastener selection:

Base Metal	Fasteners
Aluminum	Stainless steel or aluminum alloy
Copper	Copper or brass
Stainless Steel	Stainless Steel
Terne	Cadmium plated or galvanized steel fasteners

METAL THICKNESS FOR ROOFING AND FLASHING

When using aluminum, copper, stainless steel and terne metal for various roofing and flashing applications, the thickness for each metal and application should be as shown in Table 7-7. In addition, details for execution, as recommended by SMACNA, Sheet Metal and Air Conditioning Contractors' Association,

in its *Architectural Sheet Metal Manual,* are likewise noted in Table 7-7.

PHYSICAL PROPERTIES

The tensile strength and coefficient of thermal expansion of the roofing metals are as follows:

Metal	Tensile Strength (psi)	Thermal Expansion
Aluminum	22,000	.0000129
Copper	36,000	.0000094
Stainless Steel	85,000	.0000096
Terne	45,000	.0000065
TCS	80,000	.0000096

ALUMINUM

Aluminum is an easily fabricated and silvery appearing metal. It may be left in a mill (natural) finish where it will weather to a uniform gray, or special alloys can be anodized to obtain ranges of gray, bronze, amber or black. Aluminum may also be prepainted with acrylic, vinyl, fluorocarbon or alkyd coatings. In addition textured surfaces are available, obtained by rolling or embossing.

Alloys

The alloy and temper of aluminum generally used for sheet-metal work is 3000-H14. Where anodizing is required use specific alloys 1100, alclad 3003, 3004, alclad 3004, 3005, 5005, 5050, and 5052, usually in the H14, H24, and H34 temper.

Advantages

The advantages in using aluminum are as follows:

   1. It is a lightweight, corrosion resistant metal.
   2. It will not stain adjoining surfaces.
   3. It is ductile, malleable, and easily worked.

TABLE 7-7
METAL THICKNESS FOR ROOFING AND FLASHING

Item	Copper (oz.)	Stainless (in.)	Galvanized steel (gage)	Aluminum (in.)	Terne-Coated Stainless Steel (in.)	Terne (gage)	SMACNA Manual (plate no.)
Exposed Flashings							
Cap	16	.018	26		.015		49–53
Thru-Wall	16	.018	26		.015		
Roof Projections	16	.018	26		.015		58–61
Concealed Flashings							
Heads	10	.010	28		.015		
Thru-Wall	10	.010	28		.015		46–48
Lintels	10	.010	28		.015		46–47
Roofing							
Flat Seam	16	.018	26		.015	28	84
Standing Seam	16	.018	26		.015	26 & 28	85–87
Batten Seam	16	.018	26	.032	.015	26 & 28	88–90
Copings	16	.018	24	.032			68–72
Expansion Joints	16	.018	24	.040			76–81
Gravel Stops	16	.018	26	.025	.015		36–41
Gutters	16	.018	26	.025	.015		1–23
Leaders	16	.015	26	.025	.015		25,31–34
Reglets	16	.015	26				50
Scuppers	16	.018	24	.032			26–30

Source: Copyrighted by Sheet Metal and Air Conditioning Contractors National Association (SMACNA). Used with permission.

Precautions in Use

Since aluminum has a high thermal coefficient of expansion, care must be exercised in detailing to provide for thermal movement. Large flat surfaces should be avoided to prevent waviness. Aluminum sheet metal should not be soldered; instead, mechanical fasteners or welds should be used. Where aluminum is used with other metals or is in contact with concrete or masonry the aluminum should be coated with bituminous paint to act as a dielectric separator.

Sizes

Aluminum is available in widths of 36 and 48 inches and lengths of 96 and 120 inches. It is also available in coil stock. Thickness is expressed in gage or decimals of an inch, preferably the latter.

Reference Standards

Aluminum may be specified to meet Fed. Spec. QQ-A-250 or ASTM B209.

COPPER

Copper is a ductile, malleable, easily worked metal with a characteristic bright reddish brown color. With exposure to the weather, it changes in color, slowly passing through various brown and green shades. In dry climates the green patina may never occur.

Advantages

Copper is resistant to corrosion in air and moisture. It is easy to work and join in the field. It is not corroded by masonry, concrete or stucco when flashed or embedded therein.

Precautions in Use

To prevent corrosion or galvanic action it is best not to use copper in direct contact with aluminum, zinc, or steel, unless a dielectric separator is used. Where the wash from copper metalwork would impinge on stone or concrete use lead coated copper or detail the metalwork to avoid water run-off to prevent staining.

Sizes

Copper is available in sheets, 36 inches wide by 96 to 120 inches long. Thickness of copper is expressed in ounces per square foot.

Reference Standards

Copper may be specified to meet Fed. Spec. QQ-C-576, ASTM B370 and Copper Development Association alloy 110 or 122.

Lead-Coated Copper

Lead-coated copper is copper that is coated with lead on both sides and has a characteristically gray color. It is used primarily to avoid staining of concrete, stone and stucco when flashed or embedded therein. The lead coating is applied by dipping or electrodeposition and weighs between 12 to 15 pounds per 100 square feet evenly distributed on both sides. The reference standard for lead coated copper is ASTM B101, Class A.

STAINLESS STEEL

Stainless steel is a highly durable, maintenance free, corrosion resistant metal with a silvery appearance. The 300 series containing chromium, nickel and manganese are recommended for roofing and flashing applications, using the dead soft, fully annealed types.

Advantages

Stainless steel has excellent corrosion resistance. It is self-cleaning and requires little or no maintenance. It does not stain adjacent surfaces and may be used in conjunction with concrete, masonry, and stone without danger of corrosive attack.

Precautions in Use

Upon completion of soldering operations, surfaces of stainless steel should be cleaned to remove flux and contaminants that may lead to surface corrosion. Long or large flat surfaces should be avoided to reduce waviness.

Sizes

Stainless steel is available in 30-, 36-, and 48-inch wide sheets by 96 and 120 inches long. Thickness is given in gauge or decimals.

Reference Standards

Stainless steel may be specified to meet Fed. Spec. QQ-S-766 and ASTM A167.

Alloys and Finishes

The alloys used for roofing are types 301, 302, and 316. Type 316 although expensive is recommended for highly corrosive areas such as industrial, chemical, and seacoast atmospheres.

Finishes typically used include mill finish 2D and 2B, with No. 4 used for architectural appearance.

TERNE

Terne metal consists of copper bearing steel, coated both sides with a lead-tin alloy. Coating weights are 20 pounds and 40 pounds, per 100 square feet.

Advantages

Terne metal is used primarily for sheet-metal roofing, most often flat seam, standing seam, and batten seam. When painted and maintained it is durable and has long-term permanence. It is lightweight and has a low coefficient of expansion.

Precautions in Use

Terne metal should be prime painted on both surfaces prior to installation and the exposed side painted soon after installation. It is recommended that cleats be used and that no fasteners be driven through the metal. Terne should not be in contact with aluminum, copper, or acidic materials.

Sizes

Terne metal is available in cut sheet and in rolls of varying widths and lengths, and in thickness of 30, 28, and 26 gage.

Reference Standards

Terne may be specified to meet Fed. Spec. QQ-T-191 (long terne), QQ-T-201 (roofing terne) and ASTM A308.

Terne-Coated Stainless (TCS)

TCS uses a Type 304 stainless steel sheet with coating on both surfaces of a lead-tin alloy. The material may be exposed to weather to a uniformly dark gray. Because of its core and coating, TCS is highly resistant to corrosion in severe atmospheres.

THERMAL INSULATION

The characteristics of thermal transmittance and resistance of materials are discussed in Chapter 1 under Thermal Properties.

Thermal insulation is utilized in the design of structures for two major reasons: (1) to provide a more comfortable and healthful environment for occupants, (2) to save on energy costs, since heat losses and heat gain are reduced considerably.

HEAT TRANSMISSION

Heat loss and heat gain are transferred through two mechanisms—one by infiltration, the other by transmission. Infiltration occurs through cracks around doors, windows and other openings and is best controlled by sealants and weatherstripping. Heat transmission or heat transfer takes place through roofs, walls and slabs. Slabs would include a plaza over occupied space below grade, or the underside of an elevated floor that is exposed above grade. The heat losses and heat gains through these elements are best controlled by insulation and sometimes by dead air spaces.

Most building materials are effective conductors of heat. To counterbalance heat transmission, heat insulators or resistors are required. Materials rated by their resistance to heat flow R, are good if the resistance value is high, poor if it is low. Table 7-8 illustrates the R-value of typical building materials and insulations.

TABLE 7-8
R-VALUES OF BUILDING MATERIALS AND INSULATIONS

Material	R-Value
Concrete, per inch	.08
Common brick, per inch	.2
Face brick, per inch	.11
¼ inch glass	.17
Insulating glass	1.61
Air space, 3 inches	0.91
Wood fiberboard, 1 inch	2.7
Glass fiberboard, 1 inch	3.8
Perlite board, 1 inch	2.7
Polystyrene board, 1 inch	4.2
Polyurethane board, 1 inch	6.7
Cellular glass, 1 inch	2.5

WATER VAPOR

In Chapter 1, under the heading of Water Permeability, the characteristics of water vapor, its diffusion and condensation are discussed. Since the use of insulation is on the increase due to the cost of energy, the problems of condensation and its effect on insulation and building materials pose problems that should be addressed.

Since interior spaces in buildings are kept at warmer temperatures in winter than outdoor temperatures, the moisture content of the warm air exceeds that of the exterior. Water vapor has a tendency to move from areas of higher relative humidity to that of lower humidity. If water vapor is permitted to move through insulation unimpeded, condensation will occur at the dew point somewhere within the insulation or in the building structure. When insulation absorbs moisture its R-value is diminished, thereby reducing the effectiveness of the insulation. Condensation occuring within or on other elements of the building construction may result in other deleterious effects such as corrosion, swelling, cracking and peeling.

To offset the problem of condensation, the use of vapor barriers or vapor stops depending on the perm rating desired should be considered. These barriers are normally placed on the warm side of the insulation and the joints of the barrier taped to insure a continuous, unbroken moisture stop.

ASTM C755 is a recommended practice for selection of vapor barrier for thermal insulations and describes design principles and procedures for vapor barrier selection for control of water vapor flow through thermal insulation.

TYPES OF INSULATING MATERIALS

Insulating materials may be classified as to type, which is a function of their composition (glass, plastic, rock) and their internal structure (cellular, fibrous). The types of insulation may be classified as follows:

1. *Fibrous—composed of small-diameter fibers.* Glass, rock wool, slag wool, wood, paper, and synthetic fiber are commonly used.

2. *Cellular—composed of small, individual cells separated from each other.* Glass, polystyrene, polyurethane, and urea-formaldehyde comprise this group.

3. *Granular—composed of small nodules containing air or hollow spaces.* May be produced as a loose pourable material or, as in the case of perlite combined with a binder to make a rigid material. Vermiculite, perlite, and cellulose are examples of this group.

4. *Air spaces—created by reflective or nonreflective surfaces.*

FORMS OF INSULATING MATERIALS

Insulating materials may also be classified as to physical form which is a function of the shape or application. The physical forms of insulation may be classified as follows:

1. *Rigid.* These materials are available in preformed shape as block, board, sheet, pipe covering, and curved segments produced to standard lengths, widths, and thickness. Cellular, granular and fibrous types are produced in these physical forms.

2. *Flexible and Semirigid.* These are materials with varying degrees of compressibility and flexibility available as blankets, batts, and felts and produced in sheets and rolls of varying lengths, thicknesses and widths. Cellular fibrous types are produced in these physical forms.

3. *Loose-fill.* This physical form consists of fibers, granules or nodules, poured or blown into enclosed spaces, walls, and attic floors.

4. *Foamed-in-Place.* These are materials which are available as liquid components which interact upon mixing and can be poured, frothed or sprayed in place to form rigid or semirigid foamed insulation. These materials are the cellular plastics.

INSULATING MATERIALS

Rigid

Glass: Glass fibers, felted and treated with resinous binder and compressed. These board type insulations may be faced with vapor barrier type materials having perm ratings as low as .02, or they may be left unfaced.

Wood or Cane: Felted and compressed and impregnated with a binder to resist moisture absorption; fabricated into board stock.

Perlite: Glassy volcanic rock expanded by heating; compressed into board form with a resinous binder and fibrous material.

Polystyrene: A plastic material expanded into foam by using air and fluorocarbon and molded into board form.

Polyurethane: A plastic material expanded into foam using fluorocarbon and molded into rigid boards.

Cellular Glass: Glass, ground and remelted expanded into a foam with air and produced in block form.

Flexible and SemiRigid

Glass: Glass fibers, felted and produced as batts, blankets and semirigid boards, faced or unfaced with vapor barrier materials.

Rock Wool: Slag from iron, copper, and lead produced as a mineral wool in the form of a fibrous material and felted for use as batts and blankets.

Cellulosic Material: Paper, cotton, wood fiber, or synthetic material with chemicals such as borax added to resist fire, water and vermin and produced in batt and blanket form.

Loose Fill

Vermiculite: A natural mica expanded by heat to form a lightweight pellet and used in the expanded state as loose insulating fill.

Perlite: A glassy volcanic rock expanded by heating into a granular structure and used in this state as loose insulating fill.

Polystyrene: Pellets of foamed plastic in bead form used in this state as loose insulating fill.

Foamed-in-Place

Polyurethane: A plastic material combined with foaming agents, foamed, frothed or sprayed into voids of structures or on roof surfaces. Since polyurethane is combustible, application should be confined within concrete, masonry, or plaster, except for roofs.

Urea-Formaldehyde: A urea resin formulation, foamed or sprayed into voids of structures. Water vapor permeance and water absorption resistance is poorer than most other insulations. Currently, there are questions raised concerning toxic reaction from fumes generated by some formulations.

CHARACTERISTICS OF STANDARD INSULATION PRODUCTS

Table 7-9 provides some physical characteristics of current standard insulation products.

TABLE 7-9
PHYSICAL CHARACTERISTICS OF INSULATION PRODUCTS

	Supplied Form	Fire Rating	Smoke Density/ Toxicity	"K" Factors Cured	Usable Temp. Ranges	Vapor Barrier Requirements 0 to 100°F
Plastics (Cellular)						
Polystyrene (Extruded)	Board stock	Combustible	Dense/toxic	0.18–0.26	−100 − +180°F	Joints only
Polystyrene (Bead)	Board stock	Combustible	Dense/toxic	0.24 (average)	−20 − +180°F	Joints only
Polyurethane	Board stock and Foamed-in-place	Combustible	Dense/toxic	0.16–0.21	−100 − +180°F	Joints only on boards
Urea-formaldehyde	Foamed-in-place	Noncombustible	Little/ questionable	0.18–0.20	−100 − +180°F	None required
Fibrous						
Fiberglass (rigid)	Board Stock	Noncombustible	Light/	0.24–0.27	−50 − +400°F	Required
Fiberglass (flexible)	Blanket and batts	Noncombustible	Light/nontoxic	0.25–0.29	+10 − +300°F	Required
Rockwool— Mineralwool	Board stock	Noncombustible	Light/nontoxic	0.27–0.32	+40 − +800°F	Required
Rockwool— Mineralwool	Blanket and batts	Noncombustible	Light/nontoxic	0.25–0.29	+40 − +800°F	Required
Cellular Glass						
Foamglass (rigid)	Block	Noncombustible	Light/nontoxic	0.35–0.45	−100 − +1200°F	Joints only
Loose or Fill						
Polystyrene	Pellets	Combustible	Dense/toxic	0.22–0.28	+40 − +180°F	Not applicable
Vermiculite	Granular	Noncombustible	Little/nontoxic	0.27–0.42	+40 − +1800°F	Not applicable
Perlite	Granular	Noncombustible	Little/nontoxic	0.26–0.40	+40 − +1800°F	Not applicable

PERFORMANCE REQUIREMENTS (INSULATION)

When selecting, designing, specifying and utilizing insulation, refer to Chapter 1 for information on the pertinent performance requirements, and then check for the technical aspects in this chapter. Then proceed to evaluate the material as suggested in the following example.

Structural Serviceability

When evaluating and selecting an insulation material, its location will determine whether it must be self-supporting, whether it must resist compression if exposed, and whether it may be too friable if exposed.

Fire Safety

1. Flame spread: Not to exceed 25, ASTM E84.
2. Smoke development: Not over 25, ASTM E84.
3. Fuel contribution: Not over 25, ASTM E84.
4. Toxicity: Not to exceed that of red oak.

Habitability

1. Thermal resistance should not be less and thermal conductivity should not exceed values listed in manufacturers' literature.
2. Water permeability with or without vapor barrier should not exceed 0.5 perm. Vapor barrier should not exceed 0.5 perm.

3. Insulation materials should be nontoxic, emit no odors and should be mildew and vermin resistant.

Durability

1. Material should retain its dimensional characteristics when subjected to temperature and/or moisture changes.

Compatibility

1. When exposed, coatings for finishing should be compatible with insulation.

JOINT-SEALING MATERIALS

With the advent of the aluminum curtain wall in the late 1940s, there arose a need for more sophisticated, elastic and durable joint sealing materials to glaze large lights of glass and to seal joints between large segments of building elements.

While indeed there was a tremendous development of new elastomeric compounds, gaskets, and preformed tapes, it has taken some time to sort out the various formulations and sealing materials and also the terminology to deal with these developments. The terminology is not yet clearly defined and the term sealants is used somewhat loosely to cover all of these newer joint-sealing products. To put the terminology in perspective, it would be well to define those terms that constitute joint sealing materials, as well as terms allied with the entire sealing process.

Definitions

Seal: A material placed in a joint as a barrier to prevent the passage of liquids, solids, or gases; includes sealants, preformed tapes and gaskets.

a. *sealant:* A bulk compound material that has the adhesive and cohesive properties to form a joint seal. (These are field-molded by applying liquid or mastic material in the joint).

b. *preformed tape:* A tacky, deformable solid having a preformed shape and designed for use in a joint held in compression.

c. *gasket:* A preformed deformable device in the form of a continuous shape for use in joints designed to exclude liquid or gas.

Adhesive Failure: A separation of a sealant from a substrate from any cause.

Back-Up: A material placed into a joint to control the depth of a sealant.

Backer Rod: A round back-up material, compressed in place in a joint to give the optimal shape for a sealant.

Bond Breaker: A release type of material placed in a joint to prevent adhesion of a sealant to a substrate.

Cellular Material: A rubber product containing many cells, either open, closed, or both.

Cohesive Failure: The tearing apart of a sealant internally due to loading where the adhesive characteristics of the sealant exceed the cohesive capabilities.

Compression Seal: A compartmentalized or cellular gasket which is held in a joint by compression.

Elastomer: A natural or synthetic polymer that is capable of returning to its original dimensions and shape after deformation, within a short period of time.

Joint Filler: Strips of nonextruding, resilient, preformed bituminous or nonbituminous material placed in a joint to reduce depth of joint.

Modulus: In the physical testing of rubber, the ratio of stress to strain, that is, the load in psi of initial cross-sectional area necessary to produce a required elongation; a measure of stiffness.

Open-Cell Material: Material having cells not totally enclosed by its walls, thus interconnecting with other cells or with the exterior.

Shape Factor: The relationship of depth to width of a field molded sealant.

Shore "A" Durometer Hardness: A term used to identify the relative hardness of rubber-like materials by means of a hardness gage (ASTM C661).

Stress Relaxation: Reduction in stress in a sealant due to creep under strain deformation.

JOINT SEALS

Nearly every exterior joint in a structure must be sealed to exclude the weather and the elements. In addition many interior joints require a seal, primarily to close a discontinuity between varying adjacent materials and occasionally to prevent leakage of air or liquids, depending upon the occupancy or use of the structure.

Some joints are working joints (moving joints) and others may be nonworking joints (no movement or negligible movement). Working joints are introduced to accommodate (1) thermal movement resulting in expansion and contraction, and (2) moisture movement caused by swelling and shrinking due to variations in moisture content. Nonworking joints include isolation joints and control or contraction joints. In addition to these joints, there are also exterior joints where metals butt or lap one another that require seals.

The materials encountered that form a part of a joint may include concrete, masonry, metals, glass, ceramics, wood, or plastics. The joint in question may consist of the same material or combinations of the materials listed.

Openings in joints must be sealed to create a barrier against the passage of gases, liquids or other unwanted substances. For the comfort of occupants, joints are also sealed to reduce noise infiltration and rain infiltration. Structures containing liquids such as swimming pools, tanks, and reservoirs must be sealed to prevent loss of contents through joints.

Building seals in joints must perform their prime function even though they may be subjected to expansion and contraction where butt joints are encountered or to shearing where lap joints are used. All of this must be endured along with moisture changes, temperature changes, exposure to sunlight, ozone, and sometimes corrosive environments of industry.

SEALANT JOINTS

Shape Factor

Elastomeric sealants when installed in a joint will alter their shape as the joint expands and contracts; how-

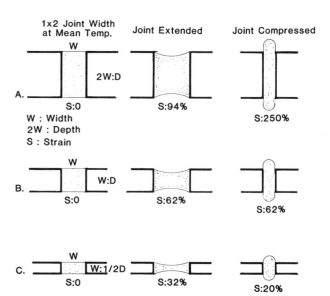

FIGURE 7-9 Joint shape factor and induced strains.

ever, their volume remains constant. The shape factor of the joint (depth to width ratio) has a critical effect on a sealant's capacity to withstand extension and compression. As the cross section of the sealant adjusts to the new size of the opening, internal strains are imposed that are often severe. The strains in the sealant and thus the adhesive and cohesive stresses developed are a critical function of the sealant shape in the joint. Actually the strain is largely determined by the depth to width ratio, and the strain on the extreme fiber is highly significant as shown in Figure 7-9.

From data developed by R. J. Schutz and reported in "Civil Engineering" ASCE, V32, No. 10, Oct. 1962 under the heading "Shape Factor in Joint Design", Figure 7-9 illustrates the critical importance and economy of using a good shape factor with elastomeric sealants.

Figure 7-9A illustrates the importance of the shape factor. A joint 1 inch wide by 2 inch deep will increase the length of the outer fiber 94% when extended ½ inch. Tests have shown that the increase in length of this outer fiber is directly proportional to the increase in strain. By reducing the depth of the joint to 1 inch as shown in Figure 7-9B, the strain is reduced to 62%. By reducing the depth of the joint to ½ inch as shown on Figure 7-9C, the strain in the outer fiber is reduced to 32%. Note also that although the joint width remains constant, the most effective depth from the point of view of strain is one-half the depth providing economy of sealant as well.

Recommended Sealant Joint Dimensions

Utilizing information obtained from the studies on shape factors and good practice from over 30 years of use, the following sealant joint configurations have been adopted by most users:

Minimum size joint; ¼ x ¼ inch
For joints up to ½ inch wide; depth = ¼ inch
For joints from ½ to 1 inch; depth = ½ the width
For joints over 1 inch; depth = ½ inch

Influence of Bond Breakers and Back-Up Materials

Function of Bond Breaker

Bond breakers which include polyethylene tape, wax paper and aluminum foil are used in a three-sided joint, generally at the bottom to insure nonadhesion of the sealant to the third side as shown in Figure 7-10A.

Function of Back-Up Material

Back-up materials such as polyethylene foam and urethane open cell foam are used to control the depth of sealant, the shape and also to support the sealant and prevent sag as shown in Figure 7-10B. The backer rod is the most common back-up material. Bond breakers are not required with the back-up materials noted herein since adhesion of sealant to back-up is

minimal. In addition these back-up materials have low surface strength and shearing capabilities so that the sealant will not be affected adversely by adhesion to the back-up. The back-up material is generally oversized and held in place by compression.

Fillers in Sealant Joints

Fillers are used primarily in horizontal work such as walks, roads, and flooring where expansion joints occur, to reduce the depth of the joint. Sometimes the filler is used to assist in making the joint and remains there. The filler is usually a preformed wood fiber, corkboard, or cellular neoprene. In addition since these fillers occur in a pedestrian or vehicular surface they also provide support for the sealant. Bitumen-saturated materials should be avoided since they may affect the sealant used. See Figure 7-11 for sealant joints with fillers.

PROPERTIES OF SEALANTS

There is no one universal sealant that possesses all of the attributes required to satisfy the requirements of all applications. Since sealants were first introduced there has been a shakedown period which allowed assessment of user needs and installation requirements and a refinement of the products initially offered and those that are on the market today.

The properties of sealants might be catalogued to include a variety of requirements. However the properties for a specific joint in a specific area of use may

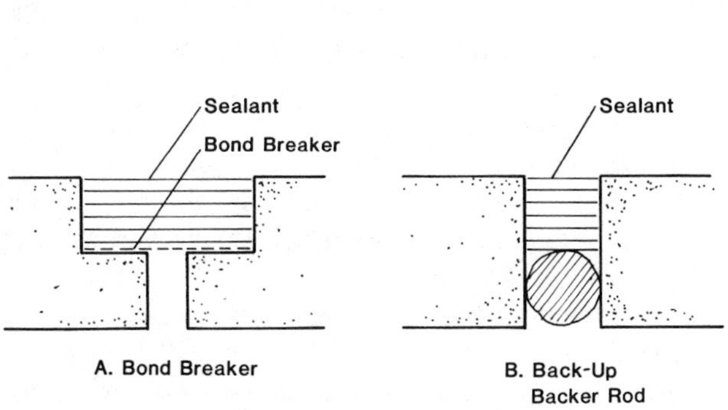

FIGURE 7-10 Function of bond breakers and back-up.

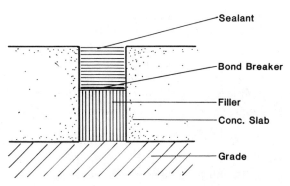

FIGURE 7-11 Fillers in sealant joints.

only encompass some of these properties. Fortunately there are a number of sealant formulations that have evolved that can to a large degree satisfy the performance requirements for specific installations.

The properties of sealants that might require investigation for a specific joint problem include the following:

Physical Property	ASTM Test
Abrasion resistance	—
Adhesion in peel	C794
Compression set resistance	—
Cyclic tension and compression	C719
Hardness	C661
Modulus of elasticity	—
Movement capability	—
Pot life	—
Resistance to heat aging	C792
Resistance to weathering	C793
Solids content	
Solvent and chemical resistance	
Stain resistance	C510
Tear resistance	—
Tensile adhesion	C719
Toxicity	—
Water resistance	—
Water immersion	—

SEALANT JOINT MOVEMENT

Movement Capability

The definition for a sealant (see Definitions herein) includes not only the highly elastomeric polymer compounds but also the old oil base and resin base caulks. To differentiate among the various types of sealants on the market today, they can be grouped on the basis of their movement capability. This distinction is perhaps the most valid since joints widths and joint spacings can be predetermined and designed and then a suitable sealant selected which has the movement capability together with those other physical properties that are pertinent for the specific joint usage. At this time there are three groups of sealants with varying movement capabilities—namely, low, medium, and high.

Low Movement

This group of sealants has a movement capability of up to ± 5% and includes the oil base and resin base caulks, and some unvulcanized butyl rubber caulks. A movement of ± 5% however means little or no movement. To begin with, any one of these materials would not be used in a joint wider than about 5/16 inch, and 5% movement would amount to 1/64 inch. That being the case, the joints in which these materials could be used safely are primarily nonworking joints such as around door and window frames.

The life expectancy of these materials is from 3 to 5 years for the poorer grades and somewhat longer for the better grades. They are generally one part products and application is relatively simple. Oil and resin base caulks may be made to meet ASTM C570 standards or Fed. Spec. TT-C-00598C and these may last up to 10 years.

Medium Movement

The sealants comprising this group have a movement capability in the range of from ± 5% to ± 12.5%, and include the following polymers; acrylics, butyls, Hypalon neoprene, and some specialty polyurethanes and polysulfides.

The acrylics (single component sealants) are produced in a number of types (emulsion, solvent and terpolymer). The emulsion types have a movement capability of about ± 7.5% but since they are emulsions they are not suited for exterior use. They may be used indoors where they are protected from water. The solvent acrylics are used for narrow joint openings, on the order of 1/8 inch or smaller, particularly for needle glazing and cracks in masonry. The solvent acrylics have movement capabilities up to ± 12.5%. The acrylic terpolymers have a movement capability of ± 12.5%, a life expectancy of about 20 years and excellent adhesive characteristics without any need for a primer.

Butyl based sealants (single component) have a variety of formulations, movement capabilities and applications. There are butyl-based caulks that have movement capabilities up to ± 7.5% and are used in nonworking joints and have also been used as acoustical sealants. Other butyl sealants have movement capabilities of up to ± 10%. There are polyisobutylene butyl sealants that are used in curtain walls for metal to metal concealed joints that remain permanently nondrying and nonskinning. They are also used for

TABLE 7-10
JOINT WIDTH VS. ± ¼ INCH MOVEMENT

% Sealant Movement Capability	Mean Joint Width (inches)	Expanded Joint (inches)	Closed Joint (inches)	Joint Depth (inches)
12½	2	2¼	1¾	½
25	1	1¼	¾	½
50	½	¾	¼	¼

bed joints for panels, rails, moldings and other two-piece metal configurations, and for heel beads for glazing.

Hypalon (chlorosulfonated polyethylene) and neoprene (polychloroprene) have movement capability up to ± 12½%, but have limited applications.

High Movement

This group of sealants, truly the elastomeric type because of their ability to deform and regain their shape rather quickly within wider ranges are the most widely used in architectural applications. They comprise the polysulfides, urethanes and silicones with movement capabilities between ± 12.5% to ±25% and more. Sealants designed to meet the requirements of ASTM C920 can be expected to perform from between 10 to 20 years or more.

This group of sealants can be used on all types of metal and glass curtain walls, between concrete panels and in masonry joints. Silicones are especially suited for glazing. Polysulfides and polyurethanes are useful for horizontal joints for traffic and pedestrian use. Polyurethane has been used successfully for traffic joints up to 12 inches wide provided metal supports are used to support the system. The silicones are primarily one part sealants, with two-part components used for specialty applications. Polysulfides and po-

lyurethanes are available in one- and two-part components and occasionally in multicomponents.

Joint Design

In the design of working joints, the factors to be considered are their spacing, size, and shape, to accommodate the computed movement. Since the ability of joints to expand and contract varies with the movement capabilities of the sealants discussed herein, the percentage of expansion required will determine the class of sealant to be used—low-, medium-, or high-performance.

It is essential, therefore, to calculate the anticipated expansion and contraction of the joint due to temperature changes. The coefficient of thermal expansion of the building material containing the joint to be sealed and the summer and winter temperature differential for the region must be known so that the width and spacing of the joint may be determined. Oftentimes, a design may be predicated on a specific panel size, such as a precast concrete panel or an aluminum panel. In such a case, the joint spacing is predetermined and only the joint width based on the temperature range of the material must be calculated.

If it is assumed that a joint will move ¼ inch in width during the contraction—expansion cycle of a

TABLE 7-11
JOINT WIDTH OF CONCRETE PANELS FOR VARYING SEALANT MOVEMENTS

Joint Movement (inches)	Sealant Movement (%)	Mean Joint Width (inches)	Expanded Joint (inches)	Closed Joint (inches)
⅛	12.5	1	1⅛	⅞
⅛	25	½	⅝	⅜
⅛	50	¼	⅜	⅛

Note: Since one must take into account the tolerances involved in precasting operations and shrinkage allowances, a mean joint width of ½ inch would be desirable.

TABLE 7-12
COEFFICIENT OF THERMAL EXPANSION OF
BUILDING MATERIALS

Material	Inches/inch °F
Aluminum	.0000129
Architectural bronze	.0000110
Brass	.0000104
Brick	.0000033
Concrete	.0000065
Copper	.0000094
Float Glass	.0000051
Granite	.0000047
Limestone	.0000044
Marble	.0000056
Stainless steel, 300 Series	.0000096
Structural Steel	.0000067
Terne	.0000065
Wood	.0000029 to .0000036

high–low temperature range, then the joint width and depth at time of installation should be as noted in Table 7-10, joint width vs. ± ¼ inch movement.

To determine the width of a joint, calculate it by using Table 7-12 for coefficients of expansion of building materials. Example: Concrete panels 12.5 feet wide are used in the design of a wall. Find the movement of the joint for a 130-degree temperature range.

Expansion of joint = .0000065 x 130 x 12.5 = 01056 ft, 0.01056 ft x 12 in/ft = 0.126 in. Width of joint based on Table 7-12 will be as shown in Table 7-11.

SEALANT MATERIALS

Sealants are the products of chemistry. Over the past 30 years since they were first introduced there have been marked changes in the various formulations and improvement in their physical properties. Chemists working in laboratories required the feedback from users as to user needs and problems of design and installation to understand and improve sealant formulation and technology. To provide detailed information on each of the types of sealants available today and their properties would render this text obsolete tomorrow, since the properties and technology are improving daily. This type of data is available from manufacturers' product sheets. However, basic information on sealants can be described herein. See Sealant Joint Movement for additional information on sealant materials.

Acrylics

These are one component materials whether they are solvent, emulsion or terpolymer types or whether they are acrylic latex caulks. Acrylics have been on the market since 1959 and are widely used around door and window frames. They have good UV resistance and excellent adhesion properties. Acrylic latex caulks can be manufactured to meet ASTM C834. Acrylics are formed by polymerization of acrylic or methacrylic acid or derivatives of either.

Butyls

These are synthetic rubbers and caulks that are formed by the polymerization of isobutylene and isoprene. Some butyls contain solvents and are therefore subject to shrinkage. They have low permeability to gases and poor resistance to abrasion. As discussed under Sealant Joint Movement, the polyisobutylenes are most useful for concealed metal to metal joints in compression.

Polysulfides

These are polymer sealants that are obtained by the reaction of sodium polysulfide and organic dichlorides. The two-component formulation was the first successful elastomeric sealant brought into the market in the early 1950s. A one component sealant was introduced in the 1970s but requires several weeks for a complete cure to take place. Polysulfides have outstanding resistance to light, oxygen, oils, and solvents.

They are good for traffic areas and where abrasion is a problem. They can also be formulated to withstand water immersion for use in swimming pools. However after 30 years of experience it has been found that their use in tropical and semitropical climates is not as successful as are the polyurethanes and silicones. Polysulfides can be manufactured to meet the requirements of ASTM C920, and Fed. Specs. TT-S-230 and TT-S-227.

Polyurethanes

These sealants are another example of the favorable elastomeric sealants, produced in the one-component, two-component and multicomponent parts. They are synthetic polymers produced by the reaction of iso-cyanates and chain-type polyols. They are used for joints in concrete panels, masonry work, and horizontal surfaces subject to traffic and abrasion. They can be manufactured to meet the requirements of ASTM C920 and Fed. Specs. TT-S-230 and TT-S-227. In addition the polyurethanes have the ability to be combined with other polymers thus permitting the development of specialty formulations providing a wider range of physical properties.

Silicones

Silicones have been on the market since the early 1950's. However because of their high price they were not utilized to a wide extent until the mid 1970s. They are essentially one part elastomeric sealants produced by the polymerization of organic siloxanes. Specialty two component silicones, especially formulated for sealing around pipe penetrations to obtain a fire rating, are also available as discussed later. Silicones offer excellent primerless adhesion to metal and glass. They have a wide temperature service range and their recovery from deformation is greater than most sealants. They offer superior resistance to UV and are not affected by ozone or oxygen. They offer movement capabilities up to ± 50%. They are not recommended for use where water immersion is expected.

SEALANTS FOR FIRE RESISTIVE JOINTS AND PENETRATIONS

A very recent development for joints and penetrations is the application of sealants and other materials for these openings to provide a measure of fire resistance when tested in accordance with ASTM E119. Since the technology is rather new, extensive use information and data is still rather meager.

Sealant manufacturers claim that when these products are used under prescribed conditions, the installation will offer protection against the passage of flames for fire ratings of 1½ to 4 hours.

SEALANT INSTALLATION

Appearance

Above all for exposed architectural applications, joint sealants should provide a good appearance, carefully tooled and weathertight joints, colored properly where required, and with adjacent surfaces unmarred by sealant smears.

Joint Preparation

Open or reduce the joint shape as required to obtain the proper width-to-depth ratio as recommended herein or as recommended by the sealant manufacturer. This may require cutting and sawing to obtain the proper width or filling with backup material to obtain the proper depth. Remove any temporary material or filler used to form the temporary sealant reservoir.

Cleaning

Joint surfaces must be clean and free of defects that would impair or inhibit adhesive bond of sealant. Dust, grease, dirt, and occasionally frost must be removed. When a coating is encountered in a joint on a steel or aluminum surface it must be compatible with the sealant and the bond developed between the sealant and the coating must not be reduced. Consult with both the sealant manufacturer and the coating manufacturer to assure proper results.

Priming

Primers are used to improve joint surfaces to obtain maximum adhesion of sealants. Depending on the surface (i.e., concrete, masonry, steel, aluminum, glass, etc.) and on the specific sealant, solicit explicit instructions from sealant manufacturer for his recommendations.

Back-Up, Fillers, and Bond Breakers

As noted before, each of these materials performs certain functions within a joint to reduce the joint depth to the optimum and to prevent three-sided adhesion.

Joint fillers in horizontal paving are often installed along with the paving and usually require only reduction in depth. See Figure 7-11. In vertical application, fillers and back-up materials are usually installed oversized so that they are in a slightly compressed mode at all times.

Back-up materials and bond breakers are usually installed just prior to application of sealant.

Weather Conditions at Time of Application

Joints are best sealed when the width approximates the mean average temperature. Since adhesive failures are more likely to occur as sealants age and are more brittle in cold weather the low side of the mean temperature would be preferable for sealant installation as opposed to the high side of the mean temperature.

Other problems associated with temperature at time of installation are (1) shortened working time for two-component materials at elevated temperatures, and (2) moisture condensation and frost on the joint substrates at lower temperatures.

Masking

Place masking tape on both sides of the joint cavity to protect adjacent surfaces from sealant smears or smudges. Install prior to priming since some primers may discolor masonry and other porous surfaces.

Applying Sealant

Install sealants as per manufacturers recommendations, preferably with a caulking gun with a nozzle whose opening may be shaped and sized to mold the required bead of sealant to fill the joint cavity. The bead of sealant should be applied without dragging, tearing, or leaving unfilled areas.

Nonsag sealants for vertical applications should be tooled to force the material into the joint and insure contact with the joint sides. The tooling also ensures a neat uniform appearance of the sealant.

Pour grade sealants for horizontal applications are best applied by filling the joint just slightly below the surface.

ASTM C962 is a standard guide for Use of Elastomeric Joint Sealants and provides detailed information on application techniques.

PREFORMED TAPES

Preformed tapes are extruded, continuous ribbon-like shapes, generally rectangular, square, round, or wedge-like sections, sometimes with embedded reinforcement, and are packaged in rolls.

They are generally the products of butyl based compounds or polybutenes. These tapes are used in permanently tacky, pressure-sensitive applications, which require compressive confinement for effective sealing, generally between metal parts. They are also used in glazing systems in surround joints as a supplementary sealant.

The tapes require no primer and are readily installed, and have good initial adhesion. The built-in-reinforcement or shim is used to control compression.

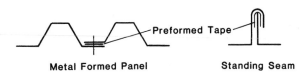

Metal Formed Panel   **Standing Seam**

FIGURE 7-12 Preformed tapes in joints.

They must be used under constant compression. Some tapes may stain porous surfaces such as concrete, masonry or stone. Since they are tacky they generally have a high dirt pick-up when exposed.

Figure 7-12 shows typical uses of preformed tapes in joints other than glazing uses.

Preformed tapes may be referenced to AAMA 804.1, Ductile, Preformed, Non-Curing Type, Back Bedding Glazing Tape and AAMA 807.1, Cured Rubber Glazing Tape.

GASKETS

The third type of joint sealing material is available in foams, open and closed cell materials and dense, rigid extrusions.

The foams or sponges are usually preformed polyurethane strips that are coated with asphalt, butylene, or some equally satisfactory waterproof material. The coated foams are sometimes used for sealing out moisture or weather. The uncoated "dry" foams are used to seal out dust and air. These gaskets must be used in compression with the degree of compression varying from 50 to 75%.

Cellular elastomeric preformed gaskets manufactured to meet ASTM C509 are furnished in a number of shapes such as rods for use as compression seals in joints, or in setting of glass and metal panels in frames. They are produced from vinyl-chloride polymers, neoprene, butyl rubbers and polyurethanes.

Dense rigid extrusions of neoprene, EPDM and PVC are produced in a variety of shapes and are utilized in glass and metal panel frames under compression for watertight joints.

Lockstrip structural gaskets are used primarily for glazing and are discussed in Chapter 8.

GLASS AND CURTAIN WALLS

GLASS

DEFINITIONS

Federal Specification DD-G-451d defines glass as:

. . . an inorganic product of fusion which has cooled to a rigid condition without crystallizing. It is typically hard and brittle and fractures in a conchoidal manner. It may be colorless or tinted and transparent to opaque. Masses or bodies of glass may be made tinted, translucent or opaque by the presence of dissolved, amorphous or crystalline material.

COMPOSITION

Glass is produced from three major ingredients—sand (silica), soda (sodium oxide), and lime (calcium oxide). About 50 other chemical compounds are also used in varying degrees to affect color, viscosity, or durability or to impart some desired physical property. An average batch contains about 70% silica sand, 13% lime, and 12% soda and small amounts of other materials.

GLASS MANUFACTURING PROCESSES

The batch containing the ingredients is fed into melting tanks or furnaces and melted at about 2800°F, and flows toward the discharge or forming end where the temperature is around 2100°F.

The reference to the manufacturing process is important since with the advent of the float glass manufacturing process and its improvement, there is essentially today only one method—the float glass process—producing flat glass architectural products. Gone are sheet or window glass and plate glass that were produced by two distinct manufacturing processes which have been replaced by the float glass process.

A minor segment of the market produces rolled glass, which consists of three types:

1. Figured or patterned glass
2. Wire glass
3. Stained or cathedral glass

FLOAT GLASS

Manufacturing Process

In 1959 Pilkington Brothers, an English company, introduced a new method for the manufacture of glass, the float glass process. In this process, the molten glass at the exit end of the furnace is poured onto a bath of liquid tin on which it floats and spreads to form a wide, flat ribbon of glass that remains untouched until it hardens.

The most striking part of this process is that, unlike the old polished plate glass manufacturing process, it produces glass having parallel surfaces, high optical quality, and fire-finished surface brilliance. It is economical in that the polished plate glass process required an investment in a huge grinding and polishing operation to obtain smooth surfaces, whereas the float process obtains these results by means of the molten tin bath.

A typical float glass furnace is about 165 feet long, 30 feet wide, and 4 feet deep. The melting tank operates continuously around the clock, 7 days a week, for about 5 to 7 years. After that, it is shut down for inspection and repair.

As the molten glass leaves the furnace and flows onto the molten tin, the speed with which it traverses the molten tin determines the thickness of the glass. Generally, the slower the speed, the thicker the resultant glass product.

The float bath of molten tin is about 150 feet long and wide enough to produce glass that is 160 inches wide.

After the glass leaves the float bath, it enters a lehr (annealing oven) where it is gradually annealed from a temperature of 1200 to 400°F to relieve internal stresses. See Figure 8-1 for a typical float glass manufacturing process.

Float Glass Products

Float glass manufacturing accounts for over 90% of the flat glass produced today. The products are clear, heat-absorbing, and tinted glasses.

Clear Glass

Clear glass is colorless and is manufactured to meet Fed. Spec. DD-G-451d, Type 1, Class 1. It is available in thicknesses for construction purposes ranging from $3/32$ inch to $1\frac{1}{4}$ inches and from 48 x 84 inches to 120 x 204 inches, depending upon the manufacturer.

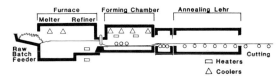

FIGURE 8-1 Advanced float glass process. Courtesy PPG Industries.

FIGURE 8-2 Lever House, New York, NY: Blue-green tinted glass used in the forerunner of curtain wall systems. Skidmore, Owings & Merrill, Architects. This illustration also appears in the color section.

Heat-Absorbing Glass

Heat-absorbing glass is intended for glazing where reduction of solar heat is required and is available in bronze, gray, and blue-green colors. The color density is related to the thickness, whereas the light transmittance is reduced with an increase in thickness. Heat-absorbing glass is manufactured to comply with Fed. Spec. DD-G-451d, Type 1, Class 2.

Heat-absorbing glass is produced by adding selected metallic oxides in small amounts to the basic glass mixture. Those oxides reduce light transmission, control solar brightness and glare, and absorb solar heat. As the glass becomes warm, it reradiates this heat, with only part of the heat directed indoors. Figure 8-2 illustrates the Lever House, which uses blue-green heat-absorbing glass.

ROLLED GLASS

Rolled glass is made by pouring molten glass from a furnace and then passing it between the rollers to obtain the required thickness. It then is annealed in a lehr and cut to required sizes. As noted previously, three types of glass are produced under this category of rolled glass: figured or patterned; wire; and stained or cathedral.

Figured (Patterned) Glass

Figured or patterned glass is produced by the use of rollers that have a pattern etched on either one or both of the rollers which imparts the pattern to one or two surfaces of the glass as it passes through the rolls. A variety of patterns is available and with differing degrees of obscuration. Figured glass can be colored, but only a limited number of colors is available. Thicknesses are usually on the order of $\frac{1}{8}$ to $\frac{7}{32}$ inch. Figured glass is made to comply with Fed. Spec. DD-G-451d, Type II, Form 3.

Wire Glass

Wire glass is produced in a manner similar to that used for figured glass, with the addition of welded wire netting or parallel wires placed in the molten glass prior to rolling. Wire glass can be obtained with a pattern (rough wire glass) or with polished faces (polished wire glass). It is made to comply with Fed. Spec. DD-G-451d, Type II, Forms 1 and 2, and is produced in thicknesses of $\frac{7}{32}$ to $\frac{3}{8}$ inch.

Wire glass is used where permissible in labeled doors and windows in accordance with the requirements of NFPA Bulletin No. 80.

MANUFACTURING MODIFICATIONS OF GLASS

While there is essentially one major method of manufacturing flat glass, there are a number of glass products that are fabricated from flat glass, including the following:

Heat-treated glass
Insulating glass
Laminated glass
Reflective glass
Nonvision glass

Heat-Treated Glass

Both heat-strengthened and tempered glass are made by a process of reheating and rapid air cooling of annealed glass. As a result of heat treating, the outer surfaces of the glass are put in compression and the central portion or core is in compensating tension.

Two methods are used in heat treating: (1) a vertical pass through the reheat furnace with the glass held by metal tongs producing tong marks; and (2) a horizontal pass on rollers. Each manufacturing method will produce a degree of bow and warp, creating some optical distortion. Limits on this bow and warp are established by Fed. Spec. DD-G-1403 B.

The heat-treating process is such that it requires all fabrication processes (cutting, drilling, edging, etc.) to be performed prior to heat treating.

Heat treatment results in increased tensile or bending strength and enables such glass to withstand greater uniform loading pressures, and solar-induced thermal stresses. As a result, for high-rise structures with varying wind-induced pressure zones, the use of the same thickness heat-treated glass in vision areas allows uniformity of light transmission, color density, and glazing detail.

Most flat glass products can be heat treated; the exceptions are wire glass and rolled glass with deep patterns.

Heat-Strengthened Glass. The heat treatment increases the resistance of this glass to thermal shock and increases its mechanical strength to 2 times that of annealed glass. When broken, heat-strengthened glass tends to remain in position in a sash opening, since its break cracks are few in number and limits relative movement. Heat-strengthened glass is usually produced in ¼ inch thickness only and does not meet the requirements for safety glazing materials.

Tempered Glass. The heat treatment increases the mechanical strength to about 4 to 5 times that of annealed glass and increases its resistance to thermal stresses. When broken, tempered glass will fracture

into many particles, minimizing the chances of injury on personal impact. Tempered glass is available in practically every thickness that is used for flat glass products and meets the requirements of Federal Standard 16 CFR 1201 for safety glazing materials.

Insulating Glass

Insulating glass units are factory-fabricated modules consisting of two panes of glass separated by a metal spacer around the perimeter, with an entrapped, sealed, desiccated air space between. Triple-glazed units utilizing a third pane and a second metal spacer are also available, primarily for use in northern climates where winter temperatures are unusually low.

Glass Elements

A wide range of glasses designed for different specific applications are utilized in insulating glass units. In addition to the typical module of clear float glass, special units for heat control, glare control, safety and acoustical requirements are produced using heat-absorbing, heat-reflecting, tempered, laminated, patterned, and wire glass.

Since the applications can be varied and the types of glass used in the assembly may be of varying types, thicknesses, sizes, and shapes, it is best to consult the manufacturer on specific requirements and recommendations before designing a specific configuration.

Components of Insulating Glass Units

The quality and life of an insulating glass unit depend on a number of factors—primarily, the type of seal, the desiccant, proper design of metal spacers, and correct corner treatment of the spacers.

Sealing Systems. Two types of seals are available—a single-seal system and a dual-seal system, as shown in Figure 8-3. The function of the seal is to prevent moisture vapor penetration (see Chapter 1), maintain a structural bond between the two panes of glass, provide adhesion to the metal spacer, and provide long-term resistance to heat, UV, and water. In the single seal type, or organic seal, polysulfide was the initial seal used, with hot melt butyls also being

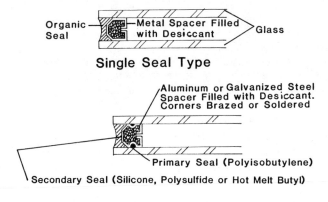

Single Seal Type

Dual Seal Type

FIGURE 8-3   Sealing systems for insulating glass.

used. In the dual seal system polyisobutylene is the primary seal which provides a superior barrier against MVT. The secondary seal which is a structural seal as well, may be polysulfide, silicone, or hot-melt butyls. Insulating units utilizing the dual seal may carry a 10 year warranty by the manufacturer.

Desiccant.   The desiccant is used to maintain a dry air space between the panes of glass. It is placed within the hollow of the metal spacer which has fine perforations or seams along the inside surface, exposing the desiccant to the entrapped air to maintain its dryness and to prevent condensation. The desiccants used are Silica-Gel and Molecular Sieve, which adsorb both water and solvent vapors from the seal.

Metal Spacers.   The metal spacers may be aluminum or galvanized steel, shaped to a rectangular or other configuration to provide the required distance between the glass panes. The corners of the spacers may be joined metallurgically by soldering, brazing, or welding, or they may be mechanically joined by an aluminum, nylon, or plastic key.

Testing and Certification

To ensure the quality and performance of insulating glass units, the Insulating Glass Certification Council (IGCC) has a program of testing and certification through specified standards. The testing standards used are outlined in ASTM E774 and ASTM E773 for Sealed Insulating Glass which classifies units into three performance categories: C, CB, or CBA, depending upon the durability of these units in the accelerated testing, with the CBA designation being the highest.

Laminated Glass

Laminated glass consists of a combination of two or more panes of glass with a layer of transparent plastic sandwiched between the panes under heat and pressure to form a single laminated unit having certain characteristics depending upon its intended use.

The plastic used is plasticized polyvinyl butyral, which may be as little as 0.015 inch thick to as much as 0.090 inch or more, as required for a specific use. The vinyl may be colorless for general use or pigmented when used for solar control.

The introduction of the plastic interlayer produces a unit that will prevent sharp fragments from shattering and flying about when it is subjected to a sharp impact and breaks since the glass adheres to the vinyl interlayer. The glass used in its lamination may consist of clear, tinted, tempered, reflective and wire glass, and the result is an adaptable, high-performance glazing material that is utilized in the following situations:

Safety glazing
Security glazing
Solar control
Sound control

Safety Glazing

The use of laminated safety glass minimizes the risk of injury from glass breakage or accidental impact. It is useful for entrance doors, sliding doors, shower enclosures, skylight, and sloped glazing to name but a few applications.

Hazardous glazing applications are identified by the Consumer Product Safety Commission in its performance standard 16 CFR 1201. Depending upon the degree of safety required, the glass components may be annealed or heat treated and the vinyl interlayer may be as little as 0.015 inch thick for Category 1 of Federal Standard 16 CFR 1201, or as much as 0.06 inch thick for overhead glazing.

Security Glazing

Security glazing is used, when properly designed, for three different levels of security as follows:

Burglary resistant

Institutional security

Bullet resistant

Burglary-resistant laminated glass: Combinations have been developed to meet Underwriter's Laboratories Standard UL972, which provides for a number of different levels of security and differing test procedures.

Institutional security laminated glass: Utilizes multiple panes of glass and alternate plies of vinyl interlayer from 0.060 to 0.090 inch thick which are fabricated for such installations as "barless" jails to provide unobstructed vision, detention centers, prisons, and mental health facilities.

Bullet-resistant glass: This glass is manufactured to meet Underwriter's Laboratories Standard UL 752 for a multiple laminated product ranging in thickness from 1³⁄₁₆ to 2 inches and suitable for use against small handguns to high-power rifles.

Solar Control

Laminated glass products utilizing pigmented interlayers of polyvinyl butyral can reduce solar energy transmission, control glare and brightness, and screen out UV, providing both utilitarian and aesthetic qualities.

Sound Control

The use of the plastic interlayer in a laminated glass unit provides a damping characteristic which enhances the acoustic performance of these units, as compared to monolithic glass and insulating glass. Table 8-1 shows typical sound transmission class values (STC) for representative glass thicknesses.

Reflective Glass

Reflective glasses have a transparent, thin metal or metal oxide coating deposited on one surface. These glasses owe their popularity to several factors: aesthetic appeal; solar control resulting in energy savings; and occupant comfort.

Since the metallic film acts as a one-way mirror,

TABLE 8-1
ACOUSTICAL VALUES OF GLASS

Thickness	STC Value
⅛ inch float	23
¼ inch float	28
½ inch float	31
1 inch insulating	32
¼ inch laminated	34
⅜ inch laminated	36
½ inch laminated	37
⅝ inch laminated	38

a person on the exterior has difficulty looking in during the day. However, at night with interior lights on, an occupant cannot see out but anyone on the exterior may see in. The metallic film reflects sunlight and reduces heat gain markedly.

Methods of Coating Deposition

The coatings are applied by three different methods and the deposition method dictates whether the coating may be exposed to the exterior, protected by placing it on the inside of an insulating unit or in a laminate, or exposed on the interior. The methods of coating deposition are:

1. *Wet Chemical Deposition.* In this process a metallic precipitation is formed on the glass surface by a chemical reaction. Since coatings formed by this process are fragile, these coatings on glass are used in insulating glass and in laminated glass, thereby protecting the coated surface.

2. *Pyrolitic Deposition.* A fine spray of metallic oxide coatings is applied to the surface of heated glass (usually tinted), which becomes fused to the glass surface. These glasses may be heat tempered and may be exposed to the weather.

3. *Vacuum Deposition.* This process utilizes a vacuum chamber, an inert gas, and electrical energy whereby metallic ions are impinged onto the glass surface, creating thin metallic film. These glasses may be used monolithically or in insulated units.

Glass coated by the wet or vacuum process cannot be heat treated after the deposition process, since those coatings would be destroyed. These coatings may, however, be applied to glass that already has been heat treated. In some instances, the coatings make clear annealed glass so highly heat absorbent that heat treatment (strengthening or tempering) is required to increase resistance to thermal stresses.

Colors and Properties of Reflective Glasses

Metallic or metal oxide coatings may be applied to both clear and tinted glasses. This in turn provides for a variety of colors and reflectances, ranging from blues to blue-greens, greens, silver, gold, bronze, and gray.

The range for transmittance, heat-gain, reflectance, and color when utilizing insulating units of varying combinations becomes so complex that manufacturers should be consulted to obtain the most recent data with respect to these values. In addition, the development of new metallic coatings and processes is proceeding at such an unprecedented rate that it would be impossible for a textbook to stay abreast of these developments.

To understand and appreciate the various properties of reflective insulating units, one must be cognizant of the glass used and the surface on which the coating is applied. Figure 8-4 shows a typical reflective insulating unit. When reviewing manufacturer's literature or discussing these units with a manufacturer, knowing the configuration of the unit as shown in Figure 8-4 will help the designer in making a selection.

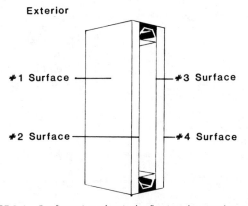

FIGURE 8-4 Configuration of typical reflective glass insulating unit. Courtesy PPG Industries.

Nonvision glass

When glass is used in a facade to hide structural members and to continue the visual effect of the vision glass both in color and textural quality, either spandrel glass or opaque-clad glass panels may be utilized.

Spandrel Glass

Spandrel glass is a heat-strengthened, ¼-inch thick glass with a ceramic frit color fused permanently to the back surface. It may also have insulation that has been factory applied to the back side with an aluminum foil vapor barrier. There is a limitless variety of colors available that will match the vision glass used adjacent to it.

Reflective glass (heat treated) with a factory-applied opacifier to create opaqueness may also be used for spandrel areas for single-glazed situations. Insulating units of reflective glass with a ceramic frit on the inner light may also be employed in spandrel areas.

Opaque-Clad Glass Panels

Glass panels with a ceramic enamel coating fired on the exposed exterior surface are produced to be utilized in a manner similar to spandrel panels. The ceramic enamel coating applied to ¼-inch thick, heat-treated glass is then coated with a protective overspray that permits standard glass-cleaning procedures. The enamel coating is available in a number of standard colors. Factory-applied fiberglass insulation is also available where required, applied to the back of the panel.

GLASS QUALITY STANDARDS

Flat Glass

Flat glass encompasses float products (clear, heat-absorbing, and tinted) and rolled products (patterned or figured and wired). Flat glass is manufactured in accordance with Fed. Spec. DD-G-451D, which is the current standard for thickness, dimensional tolerance, and other characteristics.

Heat-Treated Glass

Heat-strengthened glass, tempered glass, and ceramic-coated spandrel glass are all governed and manufactured to standards established by Fed. Spec. DD-G-1403B.

Insulating Glass Units

ASTM has established standard test methods that comprise a series of tests for insulating glass units. These standards are as follows:

E-774   Sealed Insulating Glass Units

E-546   Standard Test Method for Frost Point for Sealed Insulating Glass Units

E-576   Standard Test Method for Dew/Frost Point of Sealed Insulating Glass Units in Vertical Position

E-774   Seal Durability of Sealed Insulating Glass Units

Mirrors

Fed. Spec. DD-G-451D establishes standards for the thickness, dimensional tolerances and qualities of flat glass used for mirrors. Fed. Spec. DD-M-411 prescribes the standards for mirrors and mirror frames.

Safety Glazing

A standard promulgated by the Consumer Product Safety Commission in July 1977, (16CFR 1201) establishes standards for such items as storm doors, sliding doors, and shower doors for safety glazing for residential use, recreational buildings, schools, and public buildings. ANSI Z97.1, published by American National Standards Institute, also establishes criteria for the type of glass to be used for safety glazing and the test methods to measure effectiveness.

GLASS THICKNESS AND WIND LOADS

The major factor in the determination of glass thickness is wind pressure generated by wind velocity. Because glass is a brittle material with no specific yield point, its strength is highly variable. Tests on identical lights will show failure pressures differing sometimes by as much as 3 to 1. It is therefore necessary to express glass strength on a statistical basis.

Probability of Failure

Architects, engineers, and regulatory authorities must select a safety factor (probability of failure) for glass design that is related to public safety, performance, and economics. By selecting an appropriate safety factor, risks of breakage can be minimized but not eliminated. Table 8-2 shows the relationship of safety factors to the statistical probability of failure when a large number of lights is considered.

Generally, a design factor of 2.5 has been utilized by architects and engineers and also by the major building code authorities. These requirements may be adequate for many structures. However, for tall structures, unusually shaped buildings, and buildings where the surroundings may create unusual wind patterns, it is suggested that wind tunnel tests be run. See Metal Curtain Walls in this Chapter.

TABLE 8-2
STATISTICAL PROBABILITY OF FAILURE OF ANNEALED GLASS

Safety Factor	Probable Number of Lights[a] that Will Break at Initial Occurrence of Design Load (of Each 1000 Loaded)
1	500
2	22
2.5	8
3	4
4	2
5	1

[a]Rectangular lights adequately supported on four sides in a weathertight rabbet and assuming statistically normal strength distribution, and a coefficient of variation of 25%.

TABLE 8-3
STRENGTH OF OTHER GLASSES COMPARED TO ANNEALED GLASS

Type of Glass	Multiplying Factor
Tempered glass	4.0
Heat-strengthened glass	2.0
Insulating glass	1.5
Annealed glass	1.0
Laminated glass	0.6
Wired glass	0.5

The designer who feels that a design factor of 2.5 is inadequate may make adjustments based on personal experience by utilizing the following formula:

$$\text{Design Load} = \frac{\text{Chosen Design Factor}}{2.5} \times \text{Actual Design Load}$$

To determine the strength of other than annealed glass held four sides use the appropriate multiplying factor shown in Table 8-3. For example, if the allowable uniform load for a light of ¼ inch annealed glass is 30 psf, a light of ¼ inch tempered glass of the same size should be expected to withstand a uniform load of 120 psf and a similar light of wire glass a uniform load of 15 psf.

It should be noted, however, that while tempered glass is stronger than the same thickness and size of annealed glass, it has the same rigidity. A thicker light of tempered glass would be required in the same opening for the higher wind load to limit deflection.

Wind Load

Wind velocity pressures for a given locale are based on measurements of wind speeds developed by the U.S. Weather Bureau taken over many years at 129 airport locations. Designers may determine maximum wind velocities by referring to wind velocity maps, usually employing a 50-year recurrence map as shown in Figure 8-5. The data provided is assumed to be at an elevation of 33 feet or 10 meters.

Measurements obtained in the collection of wind data are known as "fastest mile of wind" since the procedure involves the measurement of the time required for a mile-long sample of air to pass a fixed point.

The relationship between wind velocity and wind pressure is contained in Ensewiler's Formula where $P = 0.00256\ V^2$.

P = wind pressure (psf)
V = wind velocity (mph)

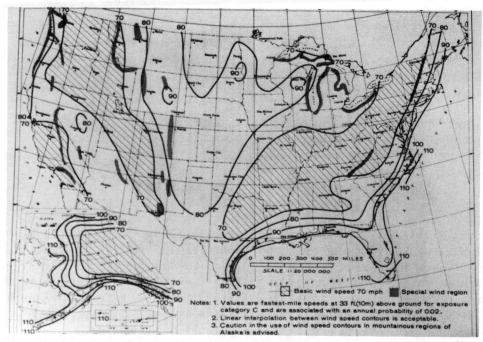

FIGURE 8-5 Basic wind speed map of United States. (This material is reproduced with permission from American National Standard A58.1 Minimum Design Loads for Buildings & Structures, copyright 1982 by the American National Standards Institute. Copies of this standard may be purchased from the American National Standards Institute at 1430 Broadway, New York, NY 10018.)

TABLE 8-4
WIND VELOCITY VS. WIND PRESSURE

Wind Velocity (speed) (mph)	Wind Pressure (psf)
70	12.5
80	16.4
90	20.7
100	25.6
110	31.0
120	36.9

When reviewing the wind velocity map (Figure 8-5), note that the wind speeds vary from 70 to 110 mph. Converting the wind speeds at 33 feet above ground using Ensewiler's Formula results in the velocity pressures shown in Table 8-4.

Wind velocity pressure should not be confused with *design pressure,* which takes several other factors into consideration and which is dealt with in more detail under the heading of Metal Curtain Walls in this Chapter. It has also been known that wind velocity increases with height above ground due to several factors, and that the type of exposure has a bearing on the wind velocity. See Curtain Walls for effect of exposures on wind design pressures.

Glass Thicknesses

The selection of glass thicknesses for low-rise to mid-rise typical rectangular-shaped buildings may be determined from charts furnished by the major manufacturers of glass. By determining the wind velocity from the wind speed map (Figure 8-5) and by using the required wind loads of the local building code, the designer may then determine the resultant *wind velocity pressures* and utilize the manufacturers charts to find the glass thicknesses for the corresponding wind velocities.

Again, the architect should be cautioned to consult with the structural engineer and with curtain wall consultants whenever unusual shapes or conditions might affect the choice of the proper *design pressure.* Also see Metal Curtain Walls in this Chapter for additional guidelines on determining *design pressures.*

It should also be noted that the wind load charts of the various manufacturers vary. In addition, the continuing use of large lights of glass all over the world in various climes is adding to the statistical information and changing the standards used 15 and 20 years ago.

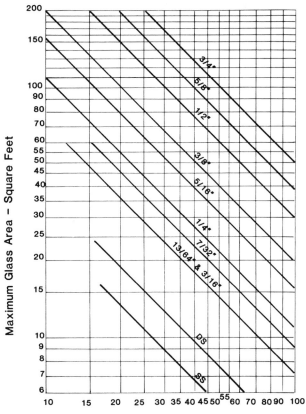

Design Wind Load – Pounds Per Square Foot
Design Factor : 2.5
FIGURE 8-6 Wind load chart.

See Figure 8-6 for the wind load chart used in Federal Construction Guide Specification, Section 08810, Glass and Glazing, dated August 1972.

SAFETY GLAZING

Standards

ANSI Standard Z97.1, Safety Performance and Methods of Test for Safety Glazing Material, Used in Buildings, sets forth the safety requirements and test methods for safety glazing materials. Its main objective is to set standards for the use of glass or plastics in hazardous locations so that they will break safely when subjected to human impact.

The Consumer Product Safety Commission on July 6, 1977, promulgated 16 CFR 1201, Safety Standard for Architectural Glazing Materials, which sets certain standards for glazing materials in residential and recreational areas, schools, or other buildings open to the public. This preempted the codes of states or political subdivisions dealing with safety glazing. This standard requires the use of certified glazing products for sliding patio doors, storm doors, interior and exterior doors, and shower and tub enclosures. Deleted from this standard in August, 1981, was a requirement for safety glazing of large panels near doors and walkways. This latter requirement is being met by the model building codes and by ANSI Z97.1.

Safety Glazing Materials

Fully Tempered Glass. For vertical applications, this type of glass qualifies as a safety glazing material under the criteria of 16 CFR 1201 and ANSI Z97.1, when so labeled and certified. Because of the manner in which it breaks, it may not be appropriate for use in skylights unlesss a wire mesh screen is utilized immediately below it.

Polished Wire Glass. Meets ANSI Z97.1 for a safety glazing material but not 16 CFR 1201. There is a risk of lacerative and puncture type injuries from the broken glass fragments as well as from protruding wires.

Laminated Glass. Meets ANSI Z91.1 and 16 CFR 1201 as a safety glazing material. When broken, glass fragments tend to adhere to the plastic interlayer.

GLAZING

Glazing may be defined as the installation or securing of glass in an opening. Glazing materials are the sealants, tapes, and/or gaskets used to provide a weathertight joint between the glass and the surround. Glazing accessories comprise the blocking, shims, spacers, clips, points, and so forth used to position and center the glass in the surround and prevent the glass from touching the frame.

Glazing, glazing design, and glazing systems deal with the installation of glass in openings utilizing varying requirements of the joint materials, joint configurations, and unusual systems and applications.

GLAZING MATERIALS

For architectural applications, glazing materials are used in wet glazing, dry glazing, and combination wet/dry glazing. In addition there are specialized glazing systems utilizing structural gaskets, butt-joint glazing, stopless or structural sealant glazing, suspended glazing, and sloped glazing.

Wet Glazing Materials

Three types of wet glazing materials are available: bulk sealants; preformed tapes; and oil-base and resin-base caulks.

Sealants: The sealants generally used for wet glazing are the curing type acrylics, polysulfides, silicones, and urethanes where exposed to the atmosphere. Noncuring, nonskinning polyisobutylene is recommended for heel beads (the material applied at the base of the channel, after setting glass and before installation of removable stop to prevent leakage past the stop). See Chapter 7, Joint Sealing Materials, for specific data on sealants.

Preformed Tapes: Butyl or polybutene based materials incorporating an integral, continuous shim as discussed in Chapter 7, Joint Sealing Materials.

Caulks: Oil-base and resin-base caulks are not especially effective for glazing medium- to large-size lights. The oils and solvents in these compounds are incompatible with most elastomeric sealants and preformed tapes; they have no place in large scale architectural applications.

Dry Glazing Materials

Dry glazing materials are preformed in a variety of shapes and comprise both soft, closed celled gaskets of PVC, neoprene, EPDM, and butyl; and firm, dense extrusions of neoprene, EPDM and PVC. See Chapter 7, Joint Sealing Materials for gaskets.

Generally for dry glazing systems, the soft closed cell gasket is used on one side of the glass and the dense firm gasket is used on the other side to provide the necessary compression. Dense compression seal gaskets are made to meet the requirements of ASTM C864.

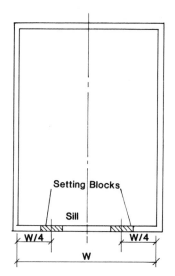

FIGURE 8-7 Setting block location.

GLAZING ACCESSORIES

Setting Blocks. For glass over 50 united inches (total of one width and one height in inches), two setting blocks are placed at the bottom of the light, spaced at the quarter points of the glass, and used to position the light in the frame. See Figure 8-7 for setting block location. Setting blocks are usually made of neoprene, with a Durometer of 80–90, and also establish the bite on the glass (see Figure 8-8). Neoprene blocks are manufactured to meet ASTM C864.

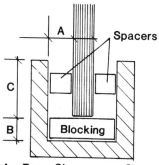

A : Face Clearance – Spacers
B : Edge Clearance – Blocking
C : Bite – Amount of Overlap
     Between Stop and Light

Check with Glass Manufacturer for Recommended Clearances for Various Glass Products and Thicknesses.

FIGURE 8-8 Typical face and edge clearance and bite.

Spacers (Shims). Small blocks of rubber, or neoprene, of 20–40 Durometer installed on each side of a light to center it in the glazing channel and maintain a uniform width of sealant bead. Used primarily with wet glazing and may be omitted where a preformed tape with an integral continuous shim is used.

Spacers are generally 1 to 3 inches long and spaced 18 to 24 inches O.C. around the perimeter of the light. See Figure 8-8 for typical face and edge clearance and bite. Spacers should be specified to meet ASTM C864.

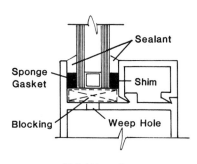

(A) Wet Glazing

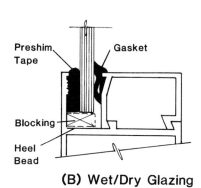

(B) Wet/Dry Glazing

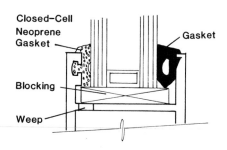

(C) Dry Glazing

FIGURE 8-9 Typical wet and dry glazing systems.

TYPICAL WET AND DRY GLAZING SYSTEMS

In the typical wet glazing system there may often be several different sealants or preformed tapes within the glazing joint. These must be checked for compatibility. In addition, if insulating glass units or laminated glass is used, the sealants used in the insulating unit and the butyl interlayer of the laminated glass must be checked for compatibility with the sealants used in the wet glazing system.

Typical wet and dry glazing systems are illustrated in Figure 8-9. For more detailed information on glazing refer to the Glazing Manual published by Flat Glass Marketing Association (FGMA).

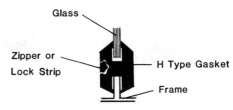

FIGURE 8-11   Typical H-type gasket.

STRUCTURAL GASKETS

Structural gaskets (often called lock-strip or zipper gaskets) are composed of solid, dense elastomeric material, primarily neoprene, and are manufactured to meet the requirements of ASTM C542. The first major building utilizing structural gaskets was the General Motors Technical Center built in 1952. An example of a project utilizing structural gaskets is shown in Figure 8-10.

Initially an H configuration structural gasket was developed primarily to hold glass or panels in an exterior frame as shown in Figure 8-11. After the gasket is installed over the frame, the lock strip or zipper is forced into the groove. As a result, a compressive force is transferred to the lips which induces a pressure to the glass and the frame.

The prime performance characteristics of the structural gasket are to provide weathertightness against water and air infiltration and structural integrity under wind loads.

A later development was the reglet or splined or grooved gasket, which was produced to be used in concrete and now adapted for metal frames, as shown in Figure 8-12. The advantage of the reglet or splined gasket is that roll-out of glass from the gasket is reduced. Consult with manufacturers of structural gaskets to obtain recommendations on allowable wind load for size and thickness of glass.

FIGURE 8-10   9 West 57 Street, New York, NY: structural gasket glazing. Skidmore, Owings & Merrill, Architects. (Photograph © Wolfgang Hoyt/ESTO)

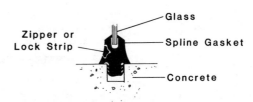

FIGURE 8-12   Typical spline gasket.

FIGURE 8-13　Interlocking gasket.

In additon to the H and reglet gasket there are special interlocking proprietary gaskets that are designed to achieve greater rollout resistance of the gasket from the frame by mechanically interlocking the gasket with the frame as shown in Figure 8-13.

The use of setting blocks, weep holes, and supplementary wet sealant application should be investigated both with the gasket manufacturer and with the glass manufacturer when structural gaskets are utilized with insulating glass and laminated glass.

Recommended installation procedures of structural gaskets are contained in ASTM C716 and C964.

BUTT JOINT GLAZING

Where architectural designs require a wide horizontal expanse of glass uninterrupted by vertical mullions, the effect can be achieved by butt glazing. The top and bottom of the glass lights are framed into a two-edge support system with wet or dry glazing systems. However, the vertical open joints between the glass lights are sealed with a silicone sealant in what is termed *butt glazing*.

Inasmuch as the butt-glazed sealant joint is utilized as a weather seal only it cannot be considered to be structural. Glass manufacturers have developed wind load charts for two-side supported glass where the span between the top and bottom frames is expressed in inches and the wind load in psf.

Attention must be paid to glass deflections in the absence of vertical supports, and this information should be obtained from glass manufacturers. Like-

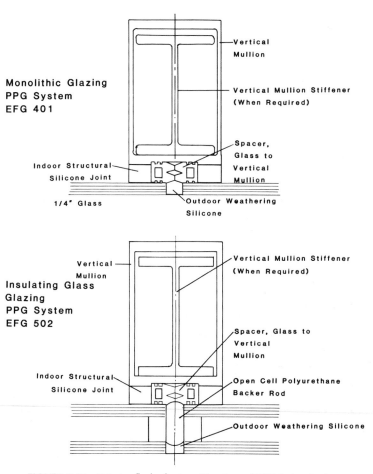

FIGURE 8-14　Exterior flush glazing. (Courtesy of PPG Industries.)

wise, building codes should be checked to ascertain whether this type of detailing is permissible under the code to be used.

Insulating units and laminated glass should not be employed unless the compatibility of the sealants is checked, since the sealants and the butyral interlayer of the laminated glass could be affected by the silicone sealant.

It is suggested that information concerning recommended joint width be obtained from the glass manufacturer for the varying glass thicknesses, when butt joint glazing is contemplated.

EXTERIOR FLUSH GLAZING

Unlike butt-joint glazing, which permits the elimination of vertical framing members but must utilize thicker glass lites since it is only two-sided supported, exterior flush glazing or stopless glazing relies on the adhesive characteristics of the sealant employed to resist negative wind loads in lieu of metal framing members. Exterior flush glazing utilizes four-sided support for the glass but has the appearance of frameless glazing. It also permits the use of insulating glass units.

In this system the vertical mullions are indoors and, in combination with structural silicone, provide four-side support for positive and negative windloads. Figure 8-14 illustrates monolithic and insulating glass exterior flush glazing details.

At present, head and sill framing is required to carry the gravity load of glass. However it is anticipated that interior framing systems will be developed that will support the glass vertically (similar to stone facades) so that the glazing systems will appear to be virtually frameless.

SPECIALIZED GLAZING SYSTEMS

Sloped Glazing

By far the most significant trend in architectural styles has been sloped glazing. Sloped glazing may be de-fined as the fenestration of enclosures that are more than 15° off the vertical, beneath which are spaces occupied by people or through which they pass. In addition local building codes should be reviewed to ascertain the legal and structural criteria pertaining to these areas.

Glass Breakage

Since glass breakage may occur in sloped glazing the designer should consider that breakage may result from any of the following causes and design accordingly:

1. Excessive wind, snow, or ice loads.
2. Inadequate or improper glazing design.
3. Thermal stresses.
4. Edge or surface damage of glass due to handling or fabrication.
5. Impact breakage due to hail or wind-borne objects.
6. Long-term weathering.

Since the most serious consequences of glass breakage is glass fallout resulting in bodily injury, it is essential that products be selected and glazing systems be designed for sloped glazing that will control this fallout.

Recommended Glazing Systems

Glass manufacturers and the Architectural Aluminum Manufacturers Association (AAMA) have developed guidelines and recommendations for sloped glazing. See AAMA TIR-A7-83, Sloped Glazing Guidelines.

Avoid the use of wire glass and fully tempered glass. Rely instead on laminated heat strengthened glass in both monolithic single glazing applications and as the bottom lite in insulating glass units. Consult the glass manufacturer and the AAMA publication for additional guidelines. See Figure 8-15 for example of sloped glazing.

Suspended Glazing

In this system a glass facade can be suspended from its top edge by hangers. These hangers clamp the top

FIGURE 8-15 First Wisconsin Plaza, Madison, WI: typical sloped glazing. Skidmore, Owings & Merrill, Architects. This illustration also appears in the color section.

FIGURE 8-16 Suspended glass glazing.

edge and the building structure carries the entire weight of the glass. In lieu of metal framing members, glass mullions or stiffeners are similarly supported, projecting as fins and secured to the suspended glass by metal patches, and sealants. The sill is designed solely to resist positive and negative wind pressures. Pilkington Glass is one of the major producers of a suspended glazing system. See Figure 8-16 for example of suspended glazing.

GLAZING CONSIDERATIONS FOR VARIOUS GLASS PRODUCTS

Insulating Glass

Weep Systems

The glazing system should be designed so that moisture or water does not accumulate and remain in the glazing rabbet, since exposure to water for prolonged periods may cause failure of the organic edge seal. Figure 8-17 shows typical weep systems for insulating glass glazing systems.

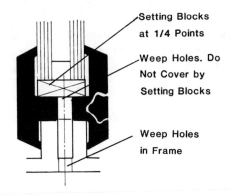

(A) Gasket Glazed System

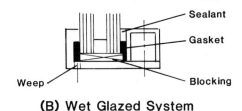

(B) Wet Glazed System

FIGURE 8-17 Weep systems for insulating glass glazing.

Sealant Compatibility

Wet glazing materials (sealants and preformed tapes) must be compatible with the organic edge seal of the insulating glass unit. Consult both the sealant manufacturer and the insulating glass unit manufacturer for information on compatibility.

Shading Patterns

When heat-absorbing or reflective glasses are used, avoid outdoor shading patterns that create double diagonal shading as shown in Figure 8-18. Also see discussion on tinted and reflective glass glazing considerations, reviewed herein.

Laminated Glass

Weep Systems

Laminated glass edges may experience delamination or haziness around the periphery when used in hot,

humid conditions over prolonged periods of time. A weep system similar to that recommended for insulating glass in Figure 8-17 should be utilized to ensure a dry environment.

Sealant Compatibility

Sealants and preformed tapes used in conjunction with laminated glass should contain no solvents or oils which would react with the butyral interlayer. Generally, 100% solid system polysulfides, silicones or butyls are suitable and compatible.

Shading Patterns

When heat-absorbing or reflective glasses are used, avoid outdoor shading patterns that create double diagonal shading (see Figure 8-18).

Heat-Absorbing and Reflective Glasses

Edge Characteristics

Heat-absorbing and reflective glasses absorb more solar energy than clear glass. Because of partial shading of portions of the glazing rabbet the exposed central portion of a light becomes hotter than the shaded edges, and will expand faster than the colder shaded edges. This differential expansion results in tension stresses at or near the edges of the light.

To ensure structural reliability, the edges should be free from scratches, vents, nips, and general edge damage. The ability of heat-absorbing and reflective glass to resist solar thermal stress breakage is determined by edge strength which is a function of how the glass is cut.

Clean-cut edges following the recommendations of the glass manufacturer should be used to assure structural reliabilty. Cuts which produce edge damage may result in thermal stress breakage. Glass 3/8 inch and thicker should be ordered with factory cut edges. For information on acceptable glass edges and borderline or defective edges see PPG Technical Service Report

Thermal Stress Factors

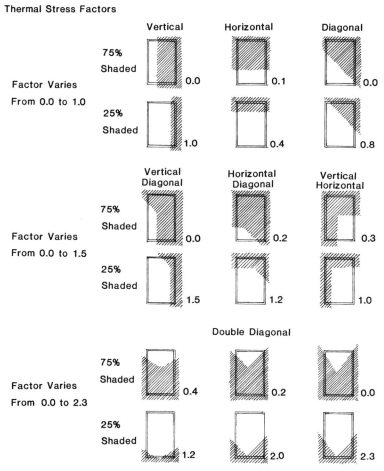

FIGURE 8-18 Outdoor shading patterns on tinted and reflective glass. Courtesy PPG Industries.

130, Installation Recommendation for Tinted and Reflective Glass.

To protect clean-cut edges from damage it is essential that the edges not be seamed, ground, polished or nipped.

Exterior Shading Patterns

Tension stresses in glass edges increase appreciably by partial outdoor shading. Roofs, balconies, overhangs, awnings, framing members, and adjacent structures may introduce shading stress. Horizontal, vertical, and diagonal shading are less severe than double diagonal shading. The most severe shading is a double diagonal shade line with a 70° included angle between shade lines intersecting at the middle of one edge. See Figure 8-18 for outdoor shading pattern.

SURFACE PROTECTION OF GLASS

WATER RUN-OFF AND GLASS DAMAGE

Glass surfaces will be affected by the run-off of rain water which is absorbed into building materials. As water enters masonry and concrete, it dissolves certain alkaline materials which may be carried down over the face of the glass. In some cases these will either stain or etch the glass.

Staining (and in some cases etching) of the glass is the result of alkalis released from concrete through water that has permeated the concrete and then leached concentrated free alkali onto the surface of the glass.

To minimize or eliminate these effects, concrete frames at window heads should be designed so that they do not splay down and back towards the glass. This particular design invites the maximum possible damage to the glass, since it creates a direct wash down the face of the glass. The introduction of an edge drip and a second drip as another line of defense should be considered (see Figure 8-19).

WEATHERING STEEL

Weathering steels (see Chapter 5) release oxides while "aging." If permitted to wash down over glass during

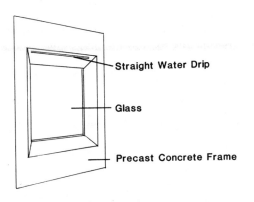

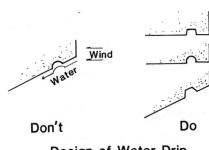

**Design of Water Drip
in Relation to Slope**

FIGURE 8-19 Eliminating water rundown on glass. (Courtesy of Prestressed Concrete Institute.)

construction without periodic washing and allowed to accumulate, the metal oxides will build up a deposit that will adhere tenaciously. This will require costly specialized cleaning to remove the residue from the glass surface.

It is recommended that a regular inspection and cleaning program be initiated early during construction to overcome the problem of residue buildup and/or staining.

IDENTIFICATION OF GLAZED OPENINGS

Identify glazed openings during construction with markers such as tapes, colorful flags or festoons that are not in contact with the glass but held in position away from the glass and attached to the framing members. Keeping these devices away from the glass prevents them from photographing through the glass when exposed to the sun.

Do not mark or coat glass with an X, floor number, or any other symbol. Such direct marking on the glass may attack the glass, be photographed into the surface, and become incapable of being eradicated.

METAL CURTAIN WALLS

Metal curtain walls are building systems made up of numerous materials described throughout this book—for example, various metals such as aluminum, steel, stainless steel, and copper alloys; glass products; insulations; coating materials; building seals; and other miscellaneous materials. Each of these materials, their properties, and performance are described throughout this book.

When these materials and products are integrated into a composite design, the ultimate metal curtain wall behaves in a specific way. It is the behavior of the wall, which is different from the sum of its parts, that is described in this Chapter.

DEFINITIONS

Curtain Wall: A nonbearing exterior building wall that carries no superimposed vertical loads.

Metal Curtain Wall: An exterior curtain wall that consists entirely or principally of metal or a combination of metal, glass, and other surfacing materials supported by or within a metal framework. See Figure 8-2.

Window Wall: A type of metal curtain wall installed between floors, typically composed of vertical and horizontal framing members, and containing fixed or operable lights or opaque panels or any combination thereof. See Figure 8-20.

FIGURE 8-20 One Liberty Plaza, New York, NY: typical example of a window wall using steel windows and a painted steel facade. Skidmore, Owings & Merrill, Architects. (Photographer, Bo Parker, New York, N.Y.)

FUNCTION OF A CURTAIN WALL

The overall function of a curtain wall (as well as any other wall) is to provide a barrier between indoor and outdoor environments so that the indoor environment can be adjusted and maintained within acceptable limits. The wall serves as the necessary barrier or filter, selectively controlling or impeding the flow inward, outward, or in both directions of those factors which affect the interior environment.

The wall acts as a barrier to wind, rain, particulate matter, insects and other elements detrimental to the interior environment. The wall also acts as a filter to selectively admit and control light, solar radiation, sound, and air.

FORCES ACTING ON THE WALL

Exterior walls are subject to the ravaging effects of nature, which include wind, temperature, rain, and sunlight. In addition the wall must be designed to deal with other factors such as fire and sound. These elements must be considered and acted upon in the design of a curtain wall.

Wind

Wind acting upon a wall produces the forces that determine the structural design of the wall, the requirements of the framing members and panels, and the thickness of glass.

Since winds also contribute to wall movement, joint seals and anchorage are also affected. Both inward and outward pressures are created by winds which in turn cause stress reversal, strains, and deformations that may permit the entry of water at affected joints.

Temperature

Temperature affects walls in two ways:
1. It induces expansion and contraction of the wall materials as a result of thermal movement;

2. It affects the interior as a result of heat movement both inward in summer (heat gain) and outward in winter (heat loss).

Rain

Water has always been one of buildings most vicious enemies. Water in the form of liquid, vapor, or condensate must be controlled by a curtain wall; otherwise, water may become trapped within the wall and cause serious damage if left undetected long enough. Freezing multiples the harmful effect of water intrusion .

Sunlight

Sunlight produces two distinct characteristics that must be dealt with: (1) solar radiation, and (2) glare and brightness. Solar radiation produces ultraviolet radiation (see Chapter 1), which has a deleterious effect on sealants, gaskets, tapes, and organic coatings unless properly formulated. Glare and brightness must be controlled through the use of tinted or reflected glasses or by interior shading devices.

Fire

Fires contribute to the products of smoke, heat, and toxic fumes which are primary causes of death in building fires. (see Chapter 1, Fire Safety.) Firestopping between the wall and the floor and at pipe and conduit penetrations significantly reduces the effects of fire.

Sound

Sounds are generated externally by any number of sources—e.g., traffic, people, and proximity to airports. Unwanted sound can be controlled and/or decibel levels can be reduced by wall insulation and the use of insulating or laminated glass.

DESIGN CONSIDERATIONS

Structural Serviceability

Since gravity loads for curtain walls are relatively light, structural design for these loadings follow well established procedures and can be dealt with quite readily. The main problem however is to understand the magnitude and nature of wind loads. The determination of *design wind pressures* often may be a complex problem.

Wind Loads

Under the heading Glass Thickness and Wind Loads in this Chapter, the conversion of wind velocity to *wind pressures* (see Table 8-4) was discussed together with the fact that height and terrain play a role and have an influence on wind *design pressures*.

Since the advent of high-rise curtain wall structures, there has been an interest in and an accumulation of knowledge on the phenomonon of wind and how it behaves. As a result, a new body of information has been developed and the most recent findings have been incorporated in ANSI Standard A58.1-1982.

Factors Influencing Wind Design Pressures

1. *Terrain.* ANSI A58.1-1982 describes four exposure categories which influence wind behavior as a result of ground surface irregularities, and constructed features. These exposures may be summarized as follows:

Exposure A—Large city centers where 50% of the existing buildings exceed 70 feet in height.
Exposure B—Urban and suburban areas, wooded areas, areas with numerous closely spaced structures no larger than single family homes.
Exposure C—Flat, open country with scattered obstructions less than 30 feet high.
Exposure D—Flat, unobstructed coastal areas exposed to winds flowing over large bodies of water.

2. *Height.* Wind velocities increase with height above ground since friction or drag at or near ground level is reduced as the height increases. ANSI A58.1

provides tables showing factors to use for various heights above grade for the different terrain exposures.

3. *Gust Effects.* Wind gusts are usually wind velocities of much shorter duration than the "fastest mile of wind" shown in Figure 8-5 but with significantly greater speeds. Gust factors increase the wind velocity and vary with height and terrain exposure, and are noted in ANSI A58.1.

4. *Pressure and Force Effects.* This is a combination of (1) external building shape and orientation; (2) internal pressures resulting from mechanical ventilation and air conditioning; and (3) the stack effect of a high-rise building. Tables giving these values are contained in ANSI A58.1.

Building Shape and Orientation

Wind forces on a building produce varying degrees of pressure on all facades, including negative or suction pressures on the leeward side. Figure 8-21 shows the air flow around a building and how it affects the air stream.

The distribution of pressure depends upon the shape of the building and how it disturbs the air flow. As the air moves over and around the building, the air flow is disturbed, creating varying positive and

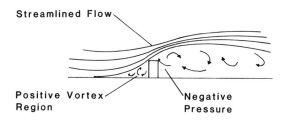

END ELEVATION

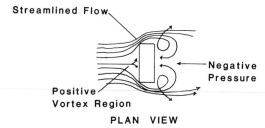

PLAN VIEW

FIGURE 8-21 Wind flow over and around a building.

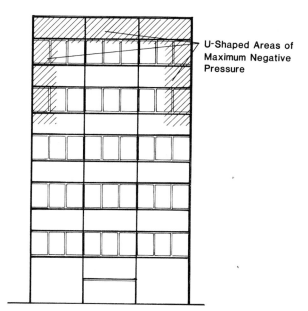

FIGURE 8-22 Areas of maximum negative pressure on a building facade.

negative pressures. As a result of wind tunnel tests and field measurements on existing buildings around the world the importance of negative or suction pressures has attained greater importance than originally conceived. Negative pressures can attain values of up to 8 to 10 times the positive pressures. For a square or rectangular building the areas of maximum negative pressures are generally at the ends and at the top of the building resulting in an inverted U shape, as shown in Figure 8-22.

Calculating Wind Loads (Design Pressures)

The *design wind loads* on structures (for purposes of determining their effect on glass and cladding) may be calculated by an analytical procedure or by actual wind tunnel tests. The analytical procedure is set forth in ANSI A58.1-1982 and involves a set of equations for velocity pressures, gust effects, and pressure and force effects.

A wind tunnel test should be conducted to ascertain the design wind pressure on the face of a structure that: (1) has an unusual shape; (2) has peculiar site conditions which would channel the winds in an unusual manner; or (3) is extremely tall.

Weathertightness

A significant publication outlining the mechanism by which water penetrates exterior walls was presented by G. K. Garden of the Canadian National Research Council in the Canadian National Digest CBD 40, entitled; Rain Penetration and Its Control. The publication states that "Rain penetration results from a combination of water on a wall, openings to permit its passage and forces to drive or draw it inwards. It can be prevented by eliminating any one of these three conditions." Leakage on a metal curtain wall therefore depends upon three factors: (1) rain water; (2) openings, or joints; and (3) a driving force such as wind. If any one of these factors is absent, no leakage will occur.

Since the materials comprising a metal and glass curtain wall are nonabsorptive, a substantial film of water can form and flow down the face of such a wall, increasing in thickness or volume as it moves downward. Should all three conditions prevail that induce leakage, considerable leakage could occur at lower levels at imperfect seals; at seals that have degraded with time; or at seals that flex and open with differential movement. One must also recognize from wind patterns shown in Figure 8-21 that the wind flow may occur sideways and upward as well as down the face of the structure and likewise carry rain water in these directions.

Two methods are in use to prevent water leakage through a wall. One that is in fairly common use in this country is "single-stage weatherproofing." The other is "two-stage weatherproofing" or "pressure-equalization" based on the "rain-screen principle" more widely used in Europe and Canada.

With the single-stage weatherproofing exterior joints are sealed. However, the designer, recognizing that water may get past these defenses at imperfect seals, degraded seals, or flexing seals, should design flashings and weep holes into the system so that water and condensation may be trapped, collected, and drained outward.

Two-stage weatherproofing or pressure equalization is based upon a treatise entitled "Curtain Walls" published by the Norwegian Research Building Institute in 1962 and authored by Øivind Birkeland in which the rain-screen principle was enunciated. The Canadian Building Digest CBD 40 "Rain Penetration and its Control" published in 1963 expanded on the research of the rain-screen principle.

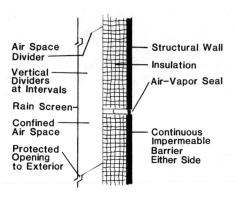

Rain Screen Principle Adapted for Metal Curtain Wall

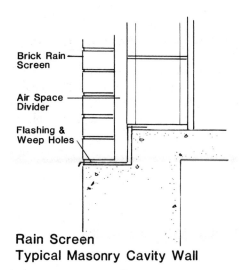

Rain Screen Typical Masonry Cavity Wall

FIGURE 8-23 Rain screen walls. (Reproduced with permission from Architectural Aluminum Manufacturing Association.)

To counteract the effects of wind-driven rain, the designer must overcome the difference in pressure that exists between the exterior face of a wall and the interior. This can be achieved by pressure equalization. Once this is accomplished, the force that drives the water through the wall is counteracted by a similar and opposite force of equal pressure. To obtain pressure equalization, an air space is provided behind the outer skin or "rain-screen," keeping the air pressure within this space equal to that on the exterior. To achieve this, air passages must be interconnected and maintained, thus maintaining equal pressures and eliminating pressure differentials which cause the driving effect of water penetration. By permitting exterior joints to be open and not sealed, pressure equalization is achieved. Seals against air leakage occur at joints, inside the wall. The masonry cavity wall has long been utilizing this principle. Figure 8-23 illustrates typical walls employing the rain screen principle.

Thermal Movement

The temperature differences that occur on curtain wall components will vary with geographical location and may be as high as 200°F and as low as 150°F. When using aluminum as framing members or panels the change in length for a 10-foot-long member may be from ¼ to ⁵⁄₁₆ inch since aluminum has a thermal coefficient of expansion of 0.0000128.

Provision must therefore be made for expansion and contraction of metal parts and for differential movement between adjacent materials with differing coefficients of thermal expansion. Sealant joints too experience expansion and contraction due to the thermal movement of the metal components. Since the temperature of the metals may be as low as 40°F at the time of sealant installation and the corresponding temperature change as much as 130°F, joint widths must be designed to accomodate the sealant's movement capability.

Thermal Performance

Controlling heat loss in winter and heat gain in summer requires the introduction of several materials into the design of the walls. Insulation materials can be readily incorporated in panels. Heat absorbing glass, reflective glass, and insulating glass units may be utilized in glass areas to improve thermal performance. The use of thermal breaks in metal framing members too will help eliminate thermal bridges that induce heat flow losses.

Sound Control

In spite of the fact that a massive wall resists sound transmission better than a lightweight wall, the metal curtain wall can be engineered to provide quiet en-

vironments through the use of sound insulation and laminated insulating glass units and by special detailing.

Fire Resistance

Firestopping in high-rise construction is an absolute necessity as an ingredient in the overall design of a metal curtain wall. Both the spread of fire and the transmission of smoke and toxic fumes to other parts of the building may occur if some adequate form of firestopping is not properly provided between the floor slab and the wall or at pipe and conduit penetrations at exterior columns and the floor.

Light Transmission

Controlling the interior environment by the selective choice of glass to reduce glare, brightness, and the effect of solar radiation is a necessary consideration in the design requirements of a metal curtain wall. Heat-absorbing glass, tinted glass, reflective glass, and insulating units, as well as laminated glass consisting of combinations of the foregoing units, can be selected to control the degree of glare, brightness, and solar heat gain through vision areas.

PERFORMANCE REQUIREMENTS

To overcome or minimize the forces acting on the curtain wall and to provide for the design considerations discussed hereinbefore, it is essential that the designs, details, and specifications include the performance requirements noted herein and in the AAMA "Metal Curtain Wall, Window, Storefront and Entrance Guide Specifications Manual."

Resistance to Wind Loads

Design the structure to withstand inward and outward wind loads. Wind load design pressures will vary with geographical location, surrounding terrain, building height, and building shape. At building corners and on the lee side, negative (outward) pressures may exceed positive (inward) pressures. Model testing in boundary layer wind tunnels should be considered for tall buildings, buildings of usual shape, or tall buildings surrounded by other buildings.

Deflections

Deflections of framing members should not exceed 1/175 of the clear span, except that when a plastered surface is involved the deflection should not exceed 1/360.

Testing

Testing for resistance to wind loads is performed utilizing ASTM E330 "Structural Performance of Exterior Curtain Walls and Doors by Uniform Static Air Pressure Difference." The test should be conducted at 1.5 times the design pressure to ensure an adequate factor of safety.

Resistance to Water Penetration

Water penetration is usually defined as the appearance of uncontrolled water other than condensation on the indoor face of the curtain wall.

Designs to control water penetration may utilize either single stage or two stage weatherproofing as described previously. Designs incorporating the rain-screen principle providing for pressure equalization, if successfully detailed, offer the best assurance against water penetration.

Testing

Two test procedures are available for determining water penetration.

The most widely used test for water penetration is ASTM E331 "Water Penetration of Exterior Windows, Curtain Walls and Doors by Uniform Static Air Pressure Difference." The air pressure difference under which the test procedure is performed must be selected by the designer. This determination is made on the basis of the geographical location since in some

FIGURE 8-24 Static test pressure assembly. (Reproduced with permission from Architectural Aluminum Manufacturers Association.)

sections of the country high winds may accompany heavy rains, whereas in other sections there may be downpours with little attendant wind speed. Usually 20% of the design wind pressure is used, with a minimum of 6.24 psf and a maximum of 12.0 psf. Water is applied by spray nozzles to the test specimen at a rate of about 5 gallons per square foot per hour, which is the equivalent of an 8-inch rainfall per hour. See Figure 8-24 for typical static pressure test assembly.

A test method to determine water penetration by dynamic air pressure has been formulated by the Architectural Aluminum Manufacturers Association AAMA 501.1-83. In this test method a wind generating device is utilized, such as an aircraft engine and propeller, capable of producing a pressure on the test specimen of 20% of the full inward design load. Water in the amount used in ASTM E331 is sprayed on the specimen. The dynamic testing creates lateral and

FIGURE 8-25 Dynamic test in progress. (Reproduced with permission from Architectural Aluminum Manufacturers Association.)

upward wind and water flow so that surface irregularities that may not cause water to infiltrate under the static test method (ASTM E331) often show up in the dynamic test. The high-frequency flutter and vibration induced by dynamic testing may also point up weaknesses in the wall that the static test does not. This test is also suggested in determining the performance of pressure equalized curtain wall systems. It is recommended that both procedures be used since each reveals different deficiencies in design and quality of work. See Figure 8-25 for a representative dynamic test procedure.

Resistance to Air Infiltration

While air infiltration is of secondary concern as compared to structural integrity and water leakage, control of air passing through a wall should be limited to a minimum amount to reduce heat loss and condensation. Good seals are important in the design of a metal curtain wall not only to eliminate water penetration but also to reduce air infiltration. Good seals also improve thermal performance and acoustical performance.

Testing

Most designers will require water penetration tests. A lesser number will specify tests for structural performance. Those failing to seek test results on air infiltration are often in the majority. However since the cost of preparing a test specimen is the major expense, an air infiltration test, once the specimen is in place, is negligible and should be included as part of the testing program.

ASTM E283 "Rate of Air Leakage through Exterior Windows, Curtain Walls and Doors" is the standard used to measure air infiltration. A pressure of l.57 psf, equal to about a 25 mile per hour wind, is generally recommended. Where high performance characteristics are expected in the wall or in operable windows, a pressure of 6.24 psf representing a 50 mph wind is utilized.

The permissible amount of air infiltration through the fixed glass areas is usually considered to be 0.06 cfm per operable square foot of area, whereas for operable sash and doors an infiltration rate of 0.25 cfm or less per linear foot of crack perimeter is acceptable.

Thermal Performance

The need for curtain wall designs which conserve energy primarily for those walls to be used in colder climates is increasingly important since many codes are incorporating larger values for thermal resistance.

Thermal tests for curtain walls are relatively expensive but may be warranted to ascertain certain characteristics such as: (1) the effectiveness of thermal breaks to find where condensation will occur—on metal or glass elements; (2) whether the details are satisfactory or require improvement to permit expansion and contraction of the elements; and (3) whether the U value computations are accurate for a composite wall.

Testing Procedures

A standardized testing procedure has not yet been formulated by ASTM or any other standards-making agency to measure air and water infiltration after cycled temperature changes, nor to observe the development of condensation.

A test method adopted by some independent researchers sets forth the following procedure:

1. Provide equipment which will maintain an interior temperature of 75°F and a RH of 30% and also lower and raise the exterior wall temperature.
2. Change the exterior wall temperature as follows:

   a. Lower the exterior temperature to 0°F and maintain this for 15 minutes.

   b. Raise the exterior temperature to 170°F for dark surfaces (130°F for light surfaces) and maintain this for 15 minutes.

3. Repeat the above procedure three times and record evidence of condensation.
4. Conduct air infiltration test (ASTM E283) and water infiltration test (ASTM E331) immediately following the three temperature change cycles. Record water and air leakage.

Sound Transmission

To control the transmission of sound through metal and glass curtain walls where sound reduction is paramount, the designer can incorporate laminated glass, sound isolation blankets, and thermal breaks in metal. Sealing of joints and cracks against water leakage likewise increases sound reduction.

Testing

Sound tests are generally conducted in the frequency range of 125 to 4000 hertz. At airports, heavily traveled highways or other major sources of unwanted sound, the performance of the wall at different frequencies may have to be investigated. STC ratings (sound transmission class) may be tested using procedures as per test method ASTM E90.

Performance of Light-Transmitting Glazed Areas

The designer should review the literature of glass manufacturers to select glass products that will satisfy the following requirements:

Light transmittance, average daylight

Thermal transmittance

Shading coefficient

Fire Resistance

While materials for a metal curtain wall may be selected which are noncombustible, have Class A flame-spread ratings, and have fire resistive ratings of 1, 2, or 3 hours, the vulnerable areas are those between the interior wall surface and the building structure, and at penetrations.

Fire-safing products have been developed in recent years for installation between the curtain wall and the building structure. Silicone products have been formulated for sealing penetrations that prevent the spread of fire and the products of combustion.

Safing is a term used to describe the fire stop between wall and floor slabs to prevent the spread of fire, smoke, and hot gases.

9

FINISHES

FLOORING MATERIALS

TYPES

The types of flooring materials are endless and continue to proliferate with the help of chemistry. Flooring materials may be categorized on the basis of the following types, which contain subtypes related to them as follows:

Carpet: Acrylic, nylon, polyester, polypropylene, and wool.

Cementitious: Monolithic concrete, concrete toppings, special shake-on (emery, quartzite, iron filings).

Clay Products: Brick, ceramic tile, quarry tile. See Chapters 2 and 4 for brick.

Resilient: Asphalt tile, cork, rubber, vinyl, and vinyl composition tile (formerly called vinyl-asbestos).

Seamless: Acrylic, epoxy, neoprene, polyester, polyurethane, polyvinylchloride.

Stone: Granite, marble, slate, travertine, sandstone, bluestone. See Chapters 2 and 4.

Terrazzo: Marble chips, granite chips in a cementitious mixture.

Wood: Hardwood strip, hardwood block (parquet), plastic-impregnated wood.

SELECTION

The selection of flooring is an important item as far as initial building costs and maintenance costs are concerned for the following reasons:

1. The cleaning and upkeep of flooring is a major item in the maintenance budget.
2. The major area of wear in a building is the floor.
3. It contributes to comfort, safety, and appearance.

There is no universal flooring material that will serve as a floor under all conditions of use. Although there are countless types of flooring, each with certain properties, the conditions of use under which the floor is expected to perform varies with the occupancy and the internal environment.

If one reviews the flooring requirements of just one type of building, a hospital, the requirements—including occupancy, use, and service conditions—would warrant a wide variety of flooring materials to meet the needs of each space: conductive flooring in operating suites; carpeting or stone flooring in lobbys; special seamless flooring in laboratories and animal rooms; resilient flooring or carpet in corridors; concrete in storage areas; carpet, terrazzo or resilient flooring in patients areas; ceramic tile in toilets, showers, and hydrotherapy areas; and wood flooring in physical therapy areas.

On what basis, then, does one make a judgment on the selection of a flooring material? By utilizing the performance characteristics outlined in Chapter 1, the essential properties of flooring material may be examined on a more rational basis and an evaluation made by a more realistic analysis.

PERFORMANCE CHARACTERISTICS

Structural Serviceability

Indentation

Of all the flooring materials only resilient flooring will yield and indent under the weight of furniture. Vinyl and rubber will have good recovery but vinyl com-

position and asphalt tile may have residual indentation.

To measure indentation there are several test methods available as follows:

Fed. Spec.   SS-T-312B
Fed. Spec.   L-F-475
MIL Spec.   D3134
ASTM   F142

Fire Safety

Flame Spread

The flame spread index for floors is generally covered by the local building code for the specific area of use. Measurement of flame spread typically required by building code is ASTM E84. Other test methods include ASTM E662, ASTM E648, ASTM E162 and MIL Spec D3134.

Smoke Development

Smoke Development (or smoke density) is a requirement usually included in local building codes and tested under ASTM E162 and E662.

Toxicity

The products of combustion, as noted in Chapter 1, may produce toxic fumes, especially among the plastic materials used in seamless or resilient flooring. Toxicity is measured by a test method cited in Chapter 1.

Habitability

Acoustic Properties

Sound transmission and sound absorption qualities are best obtained by the use of carpet, cork, and rubber flooring materials. Structure-borne sounds within a space are best controlled by the use of carpet.

Impact noise ratings (INR) for floor assemblies may be determined by methods described in FHA Bulletin No. 750.

Sound transmission Class (STC) for floor assemblies

measured by ASTM E90 may also be utilized to ascertain air-borne sound transmission losses for floor–ceiling assemblies.

Water Permeability

Cementitious, clay products, stone, and terrazzo flooring materials are for the most part unaffected by dampness and may be used on slabs on grade or below grade. However, wood and resilient type flooring products require vapor barriers, waterproofing, or mastic-type adhesives to be utilized in the details in order to overcome the effects of dampness. Manufacturers' literature must be reviewed for these latter products to ascertain the precautions necessary for proper installation under these conditions of use.

Seamless flooring and carpeting too must be protected from moisture migration when it is contemplated that they will be installed on concrete on grade or below grade. Adhesives or backing materials used in the installation of these products should be alkali resistant; otherwise, vapor barriers or waterproofing must be installed to prevent water vapor transmission.

Hygiene

Sanitation

Flooring materials that are monolithic (without seams) impervious, easily cleaned, and nonabsorbent are usually the most sanitary. Generally these include terrazzo, some seamless flooring materials and sheet vinyl. Where the joints can be waterproofed, ceramic tile and stone offer comparable degrees of sanitation.

Comfort

This characteristic is evidenced by the resiliency and shock-absorbing qualities that produce sure footedness and evenness. Comfort is obtained in the use of carpet, rubber, cork, and cushioned vinyl. Wood, vinyl and seamless are satisfactory. However, stone, clay products and terrazzo while attractive and useful for walking are not suitable for occupancies where standing is required.

Safety

Nonslip characteristics, especially where interior surfaces may be wetted accidentally or may be slightly exposed to the weather, may be measured by test methods described in Chapter 1, under Hygiene, Comfort, and Safety. Ceramic tile surfaces in shower, bath, and toilet areas may incorporate nonslip carborundum or aluminum oxide to render the surfaces nonslip.

The performance characteristic known as conductivity is specified for use in hospital operating rooms and is measured by NFPA 56A, ASTM F150 and C483. Sheet vinyl, some seamless flooring, and ceramic tile are produced to meet the specified requirement of an electrical resistance not greater than 500,000 ohms nor less than 25,000 ohms.

Durability

Resistance to Wear

This performance requirement is measured in terms of abrasion resistance, scratch resistance, scrubbing, and scuffing. Chapter 1 specifically cites examples of this requirement for resistance of flooring materials to wear under the heading Durability. It also provides test methods for abrasion resistance for various types of flooring materials. To that listing one can also include the following abrasion and scratch resistance test methods:

Fed. Test Method 141a, Method 6192
ASTM C501
Fed. Spec. L-T-345

Flooring most resistant to wear from foot traffic includes stone, clay products, cementitious materials, and terrazzo. Less durable but certainly acceptable for foot traffic would be wood and seamless flooring, whereas the remaining categories of flooring are fairly satisfactory.

Dimensional Stability

This characteristic of expansion and contraction due to thermal and moisture changes is well understood in the older flooring materials since a history of their performance is documented. However, the shinkage characteristics of flooring products which are the result of chemical formulations are less well known. When vinyl asbestos and vinyl tile were first introduced in the late 1940s, these products suffered from shrinkage due to solvent evaporation and other manufacturing defects that would leave ¼ inch joints between adjacent tiles at the end of one or two years. Some of the newer seamless floors which are primarily the products of chemistry must also be examined warily, since all of the problems associated with maintaining a dimensionally stable product may not have been worked out by the producer.

Compatibility

Some flooring materials are chosen specifically for use in special areas that may be subjected to the spillage of acids and alkalies, grease and oil, animal wastes, chemicals, or foodstuffs. Obviously the idea should be to select a flooring product for this specific service that will withstand the attack that may be induced.

In selecting such a flooring material it is essential to acquaint a prospective manufacturer with the intended use of the space and the possible spillage problems that might occur, and to obtain the necessary test data that would indicate compatibility or noncompatibility. A test method to measure chemical attack is contained in ASTM D543.

CEMENTITIOUS FLOORING

CONCRETE

Concrete floors may be monolithic where the finish is produced by surface treatment of the concrete slab. In addition, separate wearing surfaces may be placed on the structural slab using either an integral topping or a bonded topping. The integral topping is placed on a slab while it is green or before it has set. A bonded topping is applied to a slab after it has set.

Separate wearing surfaces are used to obtain ex-

posed aggregate finishes and heavy-duty wearing surfaces since these finishes can be controlled more readily. This method will be more cost-effective than attempting to obtain the same results in a monolithic slab.

Special heavy-duty floors can be produced through the use of shake-on materials such as emery, silica-quartz aggregates, and iron aggregates. For the most part these are proprietary products that are designed to harden floors for special purpose use. Manufacturers of these proprietary shake-on materials should be consulted about their products and the materials evaluated as recommended in the discussion on Performance Characteristics.

CERAMIC TILE

Ceramic tile is one of the oldest building materials, dating back over 5000 years. It is both functional and decorative. Since it is generally impervious and smooth, it is used widely in wet areas, where sanitation is important and easy maintenance is required.

To simplify the discussion of materials and installation methods, ceramic wall tile is included in this presentation.

MANUFACTURING PROCESS

Two processes are in general use in the production of tile: the dust pressed process and the plastic process.

1. *Dust Pressed Process.* The clays are finely ground and mixed with a minimum of water. The mixture is then put in filter presses where the excess water is pressed out. The resulting mixture is placed in steel dies and then fired in kilns.

2. *Plastic Process.* The clays are combined with water and mixed until a plastic consistency is reached. They are then pressed by hand or machine in dies or molds and subsequently fired in kilns.

TILE BODIES

Tile bodies may be classified on the basis of the density which is also a measure of the amount of water they will absorb. This also determines the extent of vitrification. Four classifications exist as follows:

1. *Impervious.* Tile with water absorption of 0.5% or less. These are the hardest tiles and most readily cleaned.

2. *Vitreous.* Tile with water absorption of more than 0.5%, but not more than 3%. The body density is such that dirt can easily be removed.

3. *Semivitreous.* Tile with water absorption of more than 3% but not more than 7%.

4. *Nonvitreous.* Tile with water absorption of more than 7%. Wall tile is an example of nonvitreous tile.

TILE FINISH

Unglazed tile has a homogeneous composition of clay, flint, silica, and kaoline. The color and texture is a function of that combination of ingredients which extends throughout the body.

Glazed tiles are finished by the use of ceramic materials which are applied to the body and fused on by heat. A wide variety of colors is used, some of which are plain, some mottled, some stippled, and others polychrome. The glazes may vary from a highly reflective bright glaze to a matte glaze which has very little sheen. There may also be intermediate glazes having semilustrous or satin finishes.

Special glazes are used for areas subject to abrasion or wear. Hard glazes are produced by firing at higher temperatures when the tile is to be used on floors, countertops, or kitchens. For decorative wall tile, a soft glaze is sufficient.

CERAMIC MOSAIC TILE

Ceramic mosaic tile is formed by either the dust-pressed or plastic method. It is usually ¼ to ⅜ inch thick and has a facial area of less than 6 square inches. It may be unglazed or glazed. Ceramic mosaic tiles may either be porcelain tiles or natural clay tiles. Porcelain tiles are generally made by the dust-pressed process with a composition that is dense, impervious, fine grained and smooth, with sharply formed faces. Natural clay tile is made by either the dust-pressed or plastic method from clays that produce a dense body having a distinctive, slightly textured appearance.

Unglazed ceramic mosaic tiles generally can be used for floors and walls on both the interior and the exterior. Glazed tiles may be used for both interior and exterior walls and judiciously on interior floors in dry areas not subject to wetting or heavy traffic.

Nonslip tile is produced by the addition of an abrasive admixture such as silicon carbide, or by its use in the surface or by grooves or patterns in the surface.

Mosaic tiles are usually mounted in the factory on paper, face down, in sizes of 1 x 2 feet or 2 x 2 feet to facilitate installation. The units are spaced to permit grouting between tiles.

PAVER TILE

Paver tiles are unglazed porcelain or natural clay formed by the dust-pressed method and are similar to ceramic mosaic tile in composition and physical properties. However, the major difference occurs in face size and thickness. Pavers have a facial area of 6 square inches or more and a thickness of ⅜ to ½ inch. They are generally used for interior and exterior heavy duty floors.

QUARRY TILE

Quarry tiles are generally unglazed and made from clay or shale by the plastic extrusion process. They are usually 6 square inches or more in size and from ½ to ¾ inch thick. Quarry tile is suitable for both exterior and interior use and will take moderate to heavy traffic.

Quarry tile may be glazed, but this application restricts its use for exterior floor and heavy duty flooring.

GLAZED WALL TILE

Glazed wall tile is made using a nonvitreous body and is dry pressed from soft clay. Color is achieved by adding a glaze and firing or baking to the desired finish as noted under Tile Finish.

Since the typical nonvitreous body permits moisture absorption, it is not intended for exterior use, as freezing will crack the tile body. Where glazed wall tile is required for exterior use, a vitreous body and glaze should be utilized in areas subjected to freezing temperatures.

CONDUCTIVE FLOORING

Ceramic mosaic tile is usually employed to produce conductive flooring. Finely divided carbon is added to the clay mixture to aid in the prevention of static electrical discharge where such a discharge might cause an explosion as in hospital operating rooms or laboratories.

TILE STANDARDS

STANDARDS FOR CERAMIC TILE

Tile Council of America (TCA)

TCA 137.1-1976—Recommended Standard Specifications for Ceramic Tile

ASTM

C484	Thermal Shock Resistance of Glazed Tile
C485	Measuring Warpage of Tile
C499	Facial Dimensions and Thickness of Tile
C501	Resistance to Wear
C502	Wedging of Tile
C609	Measuring Color Differences Between Tile
C648	Breaking Strength of Tile
C650	Resistance of Tile to Chemical Substances

CERAMIC TILE INSTALLATION

Ceramic tile for floors and walls can be set by two methods: (1) the older, conventional method (thick bed or mud-set) using portland cement mortar; and (2) the thin bed or thin-set method utilizing a variety of different setting materials.

The conventional full mortar bed method is particularly useful where slopes in floors are required and the mortar bed can generally achieve the slope more economically than adjusting the structural system.

The thin-set method was developed to reduce installation costs and resist certain chemical environments with specialized setting materials. Recent developments also include combination waterproofing membranes as part of the thin-set bed method. Another advantage of the thin-set bed method is that the tiles do not require soaking in water to enhance adhesion to the setting bed.

The *Handbook for Ceramic Tile Installation*, published by the Tile Council of America, provides details and covers most installation methods and conditions. It is a useful tool and can serve as a basis for developing specific details of a project with recommended specifications.

Conventional Mortar Setting Beds

Portland cement mortar setting beds are nominally ¾ inch to 1 ¼ inches thick for floors and ⅜ to ¾ inch thick on walls. Scratch coats may first be applied to walls to smooth and fill the substrate.

The installation of all types of tile (glazed wall tile, ceramic mosaic, paver, and quarry tile) in a conventional portland cement mortar bed is governed by ANSI A108-1.

Thin-Set Methods

Organic Adhesive

Organic adhesives are suitable for setting tile on horizontal and vertical surfaces on most substrates. Application of these materials is by a notched trowel to a thickness of about 1/16 inch thickness.

Organic adhesives are manufactured to meet ANSI A136.1: Type I for prolonged water resistance and

Type II for intermittent water resistance. Installation of tile in organic adhesive is governed by ANSI A108.4.

Dry-Set Mortar

Dry-set mortar is a mixture of portland cement with sand and additives which impart water retentivity. This material is used as a bond coat for setting tile and is available either factory sanded or unsanded.

Dry-set mortar may be applied to a variety of substrates but is not intended for truing or leveling these substrates. It may be applied in one layer as thin as $3/32$ inch, and also used for exterior applications.

The material for dry-set mortar conforms to ANSI A118.1; the installation of tile with this method may be performed following ANSI A108.5.

Latex–Portland Cement Mortar

This material is a mixture of portland cement, sand, and a special latex additive and is used as a bond coat in setting tile. It is used in a manner and in situations similar to dry-set mortar; however, it is less rigid than straight portland cement mortar.

Where water is anticipated after installation, as in swimming pools, gang showers, or hydrotherapy areas, it is recommended that the installation be thoroughly dry before exposure to this water. Since the latex additive varies with the manufacturer it is recommended that the manufacturer be consulted as to its proposed use and his directions followed as to the drying out period.

The latex–portland cement materials should conform to ANSI A118.4, and the installation of tile should conform to ANSI A108.5.

Epoxy Mortar

The epoxy material is essentially a 100% solid system using two or more parts to be mixed immediately before use. The epoxy setting bed is intended where chemical resistance, high bond strength, and high impact resistance are anticipated. Proper substrates are required and the intended use noted above would preclude application over gypsum board.

Epoxy materials should conform to ANSI A118.3, and the installation should comply with ANSI A108.6.

Furan Mortar

This material is a proprietary mixture consisting of a furan resin and hardener. It is intended for use where chemical resistance is paramount and is generally used on concrete, steel plate, and plywood substrates. Furan mortar sets very rapidly and is important where this requisite is essential.

Since the material is proprietary, its application should be governed by directions issued by the manufacturer. Tiles are generally prewaxed to prevent furan from adhering to the tile faces.

Conductive Dry-Set Mortar

This material is a water retentive, presanded, moderately electrically conductive portland cement mortar utilized in the setting of conductive tile. Its manufacture should conform to ANSI A118.2.

The application of conductive tile in conductive dry set mortar is governed by ANSI A108.7. Nonconductive grout is used for grouting conductive tile.

GROUTS FOR TILEWORK

Initially, grouting materials for joints in tilework were either neat portland cement for wall tile and sanded portland cement for floor tile. With the advent of thin-set installation methods and for special applications, specialized grout materials have been developed.

Commercial Sand–Portland Cement Grout

Essentially a mixture of portland cement and other ingredients that will produce a water resistant, dense, uniformly colored material. This grout is usually white for wall tile, although it may be colored. Floor grouts are usually gray. Damp curing of grout is required.

Sand–Portland Cement Grout

This mixture is essentially the conventional older portland cement–sand grout with the amount of sand added dependent upon the joint width. The grout is used for both floor and wall tile and must be damp cured.

Dry-Set Grout

A mixture having essentially the same characteristics of dry-set mortar. While damp curing is not essential, it may be desirable under very dry conditions or where strength is required.

Latex–Portland Cement Grout

Any one of previous three grouts may be modified by the use of a latex additive. The addition of the latex additive helps cure the grout, making it more resilient and less absorptive.

Mastic Grout

Mastic grout is a one-part, generally acrylic grout that does not require damp curing. It is more flexible and stain resistant than the cement grouts.

Epoxy Grout

These grouts are essentially the same materials as those used for epoxy mortars, which comply with ANSI A118.3. They provide chemical resistance, high bond strength, and impact resistance.

Furan Grout

Furan grout is the same as the furan mortar materials and offers the same properties. After the grout is applied, it cures within minutes and the tile surfaces may be open to traffic immediately. Prewaxed tile surfaces are necessary to prevent staining by the furan.

EXPANSION JOINTS

Exterior

For exterior applications, expansion joints are recommended at not more than 16 feet on centers in

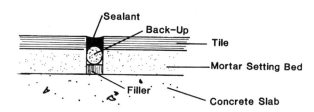

A. Conventional Mortar Setting Bed

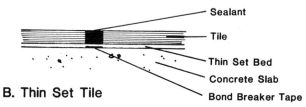

B. Thin Set Tile

FIGURE 9-1 Typical expansion joint details for tile.

both directions for horizontal and vertical surfaces; over all construction and expansion joints in the substrate; and where the substrate surface materials change. See Figure 9-1 for details.

Interior

For interior applications, expansion joints are recommended over construction or expansion joints and where the substrate materials change. For large floor areas of quarry and paver tile, expansion joints should be provided at about 24 to 36 feet on centers. See Figure 9-1 for details.

RESILIENT FLOORING

Resilient floor covering materials comprise both sheet and tile forms and include asphalt tile, cork tile, rubber tile and rubber sheet, vinyl composition tile, and vinyl tile and vinyl sheet. These materials and products have undergone considerable change in style, sizes, thicknesses, color, and pattern since these can be easily manipulated and depend on the economics of the raw materials and the marketplace. Consult current manufacturers' literature to obtain up-to-date information on sizes, thickness, colors, and costs.

Resilient floorings are used primarily indoors. They provide underfoot comfort. The foambacked vinyls offer a high degree of resiliency.

Selection of a resilient flooring material is generally based on economics, aesthetics, and the substrate and functional qualities desired.

ASPHALT TILE

Asphalt tile is made of a combination of ingredients including asphaltic and/or resinous thermoplastic binders, mineral fillers, asbestos fibers, pigments, and fillers and is formed under heat and pressure.

Asphalt tile is the least expensive of all the resilient types. It can be stained and softened by oils and animal fats and is generally brittle and hard.

CORK TILE

Cork tile consists of compressed, granulated cork bonded with a suitable thermosetting resinous binder. It is made unfaced or faced with a transparent vinyl layer. It is most comfortable, quiet and used in areas of light traffic.

RUBBER FLOORING

Rubber tile and sheet are composed of natural or reclaimed rubber, or a combination of these with reinforcing fibers, pigments, and fillers. Rubber flooring provides underfoot comfort and good resistance to indentation.

Rubber flooring may be softened by petroleum derivatives and may require buffing and waxing to retain a high gloss.

VINYL COMPOSITION TILE

Vinyl composition tile (formerly vinyl asbestos tile) is a compositon of vinyl resins, fibers, plasticizers, color pigments, and fillers, formed under heat and pressure. Asbestos fiber has been used traditionally as a filler, but its use is declining.

Vinyl composition tile is a highly versatile resilient flooring with a high resistance to abrasion, requiring little maintenance. It may be used on below grade substrates.

VINYL FLOORING

Vinyl Tile

Homogeneous vinyl tile is a blended composition of thermoplastic PVC binders, fillers, and pigments. The colorless PVC resin permits a wide range of colors. Homogeneous vinyl is a wear-resistant material having good recovery from indentation and is resistant to oil, grease, and alkalies.

Sheet Vinyl

This flooring material consists of a vinyl wearing surface compounded as described above for vinyl tile and with a backing material that may consist of felts, asbestos, foam vinyl or other type of backing. The face material provides the wearing qualities inherent in homogeneous vinyl, and the backing provides the underfoot comfort.

CONDUCTIVE RESILIENT FLOORING

Conductive resilient flooring is primarily a homogeneous vinyl with conductive carbon that is heat and pressure fused and sealed into the flooring material. A conductive epoxy adhesive is utilized with the installation to provide lateral tile-to-tile conductivity. Conductive resilient flooring installations are tested as per ASTM F140, Test for Electrical Resistance of Conductive Resilient Flooring.

INSTALLATION

Adhesives for resilient floor coverings and bases are constantly being developed and refined for the various substrates to which they are bonded. Whether the application is for wet areas, below grade areas or unusual applications, it would be prudent to consult with the manufacturer of the resilient flooring material to obtain a recommendation for the proposed application.

Materials that are available include water-base emulsions, cut-back adhesives (solutions of asphalt in hydrocarbon solvents), resinous waterproof adhesives, latex adhesives, and epoxy adhesives.

REFERENCE STANDARDS

The following is a list of standards for resilient flooring:

FEDERAL SPECIFICATIONS

SS-T-312 B

Type I	Asphalt Tile
Type II	Rubber Tile
Type III	Vinyl Tile
Type IV	Vinyl Asbestos Tile

L - F - 475 A   Vinyl Sheet

L - F - 001641   Vinyl Sheet

ASTM STANDARDS

F137   Flexibility

F142   Indentation

F510   Abrasion Resistance

F150   Electrical Conductivity

SEAMLESS FLOORING

Seamless floors are proprietary products of chemistry that utilize such resins as acrylics, epoxies, neoprenes, polyesters, polyurethanes, and polyvinyl chlorides as the synthetic resin binder together with stone chips, plastic chips, ceramic-coated granules, or graded aggregates to produce a floor covering that is unified,

unjointed, monolithic, and field applied. The flooring materials can be decorative as well as functional. By formulating specific requirements, flooring can be produced to resist chemical and industrial environments, heavy traffic, high temperature conditions, and corrosive environments.

Since seamless floors cover a wide variety of polymeric resins and aggregates, the choice of floorings is legion. In addition, there are no standards that can be used as a guide or gage to measure competing products. The user is therefore cautioned to exercise prudence in the evaluation and selection of these products. Occasionally one will find competitors using different resins and offering the same glowing advertisements for their products for similar applications. A more rational assessment can be undertaken by utilizing the evaluation method outlined in Chapter 1 and Performance Characteristic described in this Chapter.

Depending upon the specific product and the manufacturer, seamless flooring is usually applied by brush, roller, squeegee, or trowel from as little as $1/16$ to 1 inch thick. As noted previously, by adjusting the formulation various physical characteristics can be built into the product. Some are sparkproof, some are used where sanitation in a food plant is essential, some are designed for industrial floors, and some are used where an aesthetic flooring material will serve the purpose.

Chemical-resistant seamless floors utilizing epoxy resins, polyester resins, or any resin capable of forming a chemical resistance surfacing material may be evaluated for conformance to the requirements of ASTM C722.

TERRAZZO

Terrazzo is defined by the National Terrazzo and Mosaic Association (NTMA) as: "A composition material, poured in place or precast, which is used for floor and wall treatment. It consists of marble chips, seeded or unseeded, with a binder or matrix that is cementitious, noncementitious or a combination of both. The terrazzo is poured, cured and then ground and polished, or otherwise finished."

As noted in the definition, terrazzo may be used for floors and walls. It may also be used for wainscots, treads, bases, shower receptors, and other items. It is an aesthetic, functional material with an extremely durable surface available in a wide variety of color combinations, depending on the species of aggregate used in the mix.

CEMENTITIOUS TERRAZZO

Cementitious terrazzo generally consists of a cementitious underbed and a topping with divider strips. However there are variations of the above where a sand cushion is used below the underbed and also where the topping may be bonded directly to the concrete substrate. Figure 9-2 illustrates several variations of cementitious terrazzo designs.

Sand Cushion Terrazzo

Sand cushion terrazzo (see Figure 9-2 *A*) provides the best protection against cracking and transferring of building movement to the terrazzo by means of a sand cushion and an isolation membrane that divorces the system from the structural concrete base.

Bonded Terrazzo

Bonded terrazzo (see Figure 9-2 *B*) consists of a terrazzo topping installed over a cement underbed which is placed directly over a concrete slab. The underbeds are typically ¾ inch to 1 ½ inches thick.

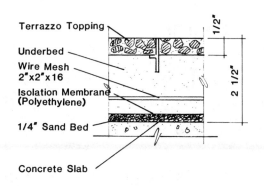

(A) Sand Cushion Terrazzo

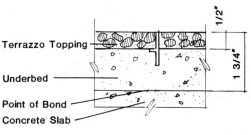

(B) Bonded to Concrete

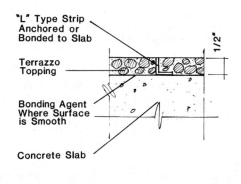

(C) Monolithic Terrazzo

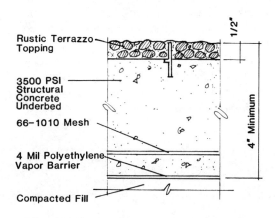

(D) Structural Terrazzo

FIGURE 9-2 Cementitious terrazzo systems. (Courtesy National Terrazzo & Mosaic Association.)

Monolithic Terrazzo

Monolithic terrazzo (see Figure 9-2 C) is the least expensive of the cementitious terrazzo systems and is used principally for large installations where the concrete slab is placed by others to a controlled height and finish.

Structural Rustic Terrazzo

Structural rustic terrazzo (see Figure 9-2 D) is intended for exterior use. In this system, the terrazzo contractor installs the structural concrete slab over a prepared base in addition to the terrazzo topping. The rustic terrazzo finish is described under Finishing Terrazzo. See Figure 2-3 for an exterior structural terrazzo installation.

CEMENTITIOUS TERRAZZO TOPPINGS

Marble Chips

Marble chips are available in a wide variety of colors and sizes. The type of marble to be used should be governed by its hardness or abrasion resistance. ASTM C241 measures the abrasion resistance of marble, and none should be used with a value below Ha 10. For exterior terrazzo, the abrasion resistance should be Ha 50 minimum. In addition no marble should be used that has a 24-hour absorption rate in excess of 0.75% for interior terrazzo and 0.25% for exterior terrazzo, since it may result in an unsightly appearance. Marble chip sizes are graded in accordance with standards adopted by the NTMA as shown in Table 9-1.

Portland Cement

To better control the ultimate topping matrix color it is best to confine the use of portland cement to white portland cement; gray portland cement is not color controlled, and variations in color may result. White

TABLE 9-1
MARBLE CHIP SIZES

Number	Passes Screen (inches)	Retained on Screen (inches)
0	$\frac{1}{8}$	$\frac{1}{16}$
1	$\frac{1}{4}$	$\frac{1}{8}$
2	$\frac{3}{8}$	$\frac{1}{4}$
3	$\frac{1}{2}$	$\frac{3}{8}$
4	$\frac{5}{8}$	$\frac{1}{2}$
5	$\frac{3}{4}$	$\frac{5}{8}$
6	$\frac{7}{8}$	$\frac{3}{4}$
7	1	$\frac{7}{8}$
8	$1\frac{1}{8}$	1

portland cement should meet the requirements of ASTM C150.

Colorants

Use alkali-resistant, nonfading color pigments.

Topping Mix

The topping mix consists of one bag of portland cement, 200 pounds of marble chips, colorant as required to produce the desired color, and 5 gallons of water.

TERRAZZO UNDERBEDS

Terrazzo underbeds are mixtures of gray portland cement and sand in the ratio of 1: 4½ plus sufficient water to provide workability at as low a slump as possible.

STRUCTURAL TERRAZZO UNDERBEDS

Structural terrazzo underbeds consist of a mixture of gray portland cement, pea or crushed gravel less than

⅜ inch diameter, sand, and water to produce a mixture with a slump less than 3 inches and a strength of 3500 psi.

DIVIDER STRIPS FOR CEMENTITIOUS TERRAZZO

Divider strips may be a white alloy of zinc, brass, stainless steel, or plastic. Thicknesses range from ⅛ to ¼ inch or more, and depths from ¾ inch to 1 ½ inches.

Expansion or control strips are sandwich-type dividers with a filler of neoprene or a removable filler to allow for installation of a bulk sealant compound.

Divider strips are utilized to control and localize any shrinkage or flexure cracks. They are also used to permit changes in the terrazzo mix design. Expansion-type dividers are used over expansion joints in the substrate and where greater movement is anticipated. Divider strips also serve as leveling guides.

FINISHING TERRAZZO

Ground and Polished Surface

Rough Grinding. Grind with 24 or finer grit stones; then follow with 80 or finer grit stones.

Grouting. Upon completion of rough grinding, apply a grout consisting of the portland cement and colorant used in the topping mix. Allow the grout to cure.

Fine Grinding. Grind the surface with 80 or finer grit stones until all grout is removed. The finished surface should show a minimum of 75 to 80% marble chips.

Rustic Terrazzo Finish

Granite or quartz chips may be substituted for marble chips for exterior rustic terrazzo. To finish, expose the aggregate by means of a pressure hose, absorbent

rolling or use of a retarder so that approximately ¹⁄₁₆ inch of the cement matrix is removed.

SEAMLESS TERRAZZO SYSTEMS

In seamless terrazzo systems, two alternative matrix materials are used. In an *acrylic* system, the latex is used to modify the portland cement so that the topping varies between ¼ and ½ inch. In the *resinous* matrix system, epoxy or polyester resins are used which may result in toppings that are ⅛ to ¼ inch thick. Consult the NTMA guide specifications for placing, finishing, and testing seamless terrazzo flooring.

Special divider strips for these thin-bed terrazzo toppings are available. These are used to simulate the appearance of standard cementitious terrazzo, not to control cracking.

Also review requirements set forth in this Chapter for seamless type floors to evaluate thin-set resin matrix terrazzo floor systems.

WOOD FLOORING

Finished wood floors used for architectural applications are generally the hardwoods that are selected for color, grain, and texture. Hardwoods that have dense, hard surfaces will withstand heavy wear and abrasion. These include, oak, maple, birch, pecan, walnut, and teak.

TYPES

Wood flooring for architectural applications is available in several forms as follows:

Strip Flooring: Long, narrow strips, with tongued and grooved edge along the sides (T & G) and on the ends (end matched).

Plank Flooring: Similar in every respect to strip floor except that it consists of wide boards.

Parquet Flooring (Thin Block Flooring): An assemblage of solid wood strips in small panel forms that varies in thickness from ⁵⁄₁₆ to ¾ inch and in face size from 6 x 6 inches to 18 x 18 inches and produces a mosaic effect in varying patterns. Sometimes, the patterns are the result of integrating varying species of wood.

Solid Block Flooring

Wood floors for heavy duty industrial and commercial applications are made in solid block form, from 2 to 4 inches thick, 3 to 4 inches wide, and 6 to 8 inches long. The species of wood used are yellow pine, and oak, edge grained. The flooring is treated against decay, vermin, and moisture with either creosote oil or pentachlorophenol preservative treatment.

Wood floors for use in certain architectural applications such as museums, lobbies, libraries, and airports are made in solid wood block form that is assembled into a strip or parquet pattern. The wood block uses douglas fir, hemlock, or pine, 1 inch to 2 ½ inches thick and 3 to 5 inches wide with an edge grain, and is preservative treated with pentachlorophenol.

Special Flooring

Specially treated wood flooring is obtained by forcing an acrylic resin into the cell structure of the wood using a vacuum/pressure cycle and then subjecting the units to gamma ray irradiation. The resultant composite is claimed to have an abrasion resistance and hardness superior to that of untreated wood. This flooring is available in strip, plank, tile, and parquet, in a variety of wood species.

GRADES

Grading for hardwood strip flooring is controlled by two associations, National Oak Flooring Manufacturers' Association (NOFMA) and Maple Flooring Manufacturers' Association (MFMA). NOFMA grades oak, beech, birch, maple, and pecan flooring. MFMA grades maple, beech, and birch flooring. The grading rules of both associations also govern sizes and lengths.

Unfinished oak is graded Clear (Plain or Quarter Sawn), Select & Better, Select, No. 1 Common, and No. 2 Common.

Beech, birch, and maple are graded First Grade, Second & Better Grade, Second Grade, Third & Better Grade, and Third Grade.

Pecan flooring is graded First Grade Red, First Grade White, First Grade, Second Grade Red, Second Grade, and Third Grade.

Prefinished oak flooring is graded Prime, Standard & Better, Standard Grade, Tavern & Better Grade, and Tavern Grade.

INSTALLATION

Parquet flooring (thin–block) and block flooring are laid in asphaltic mastic on concrete. Strip and plank flooring are laid in a variety of ways as described herein.

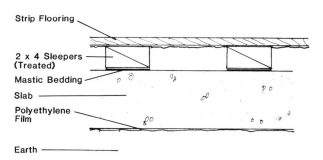

A. Strip Flooring Over Sleepers

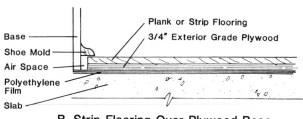

B. Strip Flooring Over Plywood Base

FIGURE 9-3 Installation of plank or strip flooring over concrete slabs.

Installation of Strip and Plank Flooring on Concrete

For slabs on grade and for slabs above grade in new construction, the use of a vapor barrier (6 mil poly-ethylene) is recommended. It should be placed on the slab with ends and edges lapped 4 inches. A wood nailing base using either plywood or sleepers is then placed over the vapor barrier and the flooring blind nailed to the nailing base as shown in Figure 9-3. Free movement of air must be provided for by allowing at least a 1-inch space around the entire perimeter. When plywood is used as a nailing base there should also be a ¼ to ½ inch space between the plywood panels.

Special Attachment and Installation Systems

For gymnasium floors, sports floors of all types, and large expanses of wood strip floors, special proprietary steel channel and clip systems and resilient neoprene pad systems are available utilizing a multitude of installation techniques.

Nailing

Strip and plank floors installed over plywood on wood sleepers must be securely and adequately blind nailed to provide a rigid installation and to prevent squeaks.

Precautions

Wood floors should be installed only after all the wet trades (masonry, plaster, tile, terrazzo) have completed their work and after the concrete slabs on which they are to be placed are dry. While the work of these wet trades is being performed, the wood flooring to be used should not be stored within the building.

CARPET

DEFINITION

Carpet: A general designation for fabric used as a floor covering of woven, knitted, or needle-tufted yarns.

FIBERS AND YARNS

Fiber

Fiber is any substance, natural or synthetic, used in thread or yarn form for processing as a textile. Pile fibers commonly used in carpet include:

    Acrylic and acrylic/modacrylic blends
    Nylon
    Olefin (Polypropylene)
    Polyester
    Wool

Yarn

Yarn is defined as a continuous strand for tufting, weaving, or knitting:

Continuous Filament Yarn: Yarn formed into a continuous strand from two or more continuous filaments.

Spun Yarn: Yarn formed from staple by spinning or twisting into a single continuous strand or yarn.

FIBERS, APPEARANCE, AND WEAR LIFE

Acrylic: Good wearing ability; low moisture absorbency leads to good soil resistance. Excellent resilience; easy to clean, good color retention; low static build-up.

Nylon: Exceptional abrasion resistance and color and texture retention. Tends to show soil more readily.

Olefin (Polypropylene):   Not attacked by mildew. Highly resistant to soiling and staining. Lacks resilience and luxurious feel underfoot. Used for indoor-outdoor carpeting.

Polyester: Excellent mildew and abrasion resistance. Good color retention.

Wool: Good wearing qualities, excellent texture retention and resiliency.

MANUFACTURING TECHNIQUES

There are several different types of carpet construction, the more important being woven, tufted, needle bonded, knitted, and loomed.

Woven carpet is produced by three basic machine techniques—namely, on Velvet, Wilton, and Axminister looms. The looms generally interweave the pile yarns and the backing yarn in one operation. These carpets are the most dimensionally stable and have good wearing qualities.

Tufted carpet today represents about 90% of all carpet manufactured. The carpet is produced by needles rather than by a weaving process. A row of needles, the width of the carpet is used to stitch the yarn into backing. The yarn is held there by a latex coating.

Needlebonded carpet is a technique used with polypropylene for the production of indoor-outdoor carpeting.

FACTORS AFFECTING CARPET QUALITY

To compare the relative quality of carpet regardless of the manner of manufacture, one need only compare (1) face weight; (2) pile height; and (3) density factors, everything else being equal. In general, the deeper, denser, and heavier the pile, the better the carpet.

Face Weight: Weight of face yarn in one square yard expressed in ounces.

Pile Height: Depth of pile between backing and top of pile, expressed in decimals of an inch. In woven carpet, pile height is given as wire size also in decimals of an inch.

Density Factor (Tufted Carpet):   This aspect is defined in terms of gauge and stitches per inch. Gauge represents the distance between rows of tufts across the width of the carpet: 1/8 inch gauge means the vertical rows of yarn across the width are 1/8 inch apart. The narrower the gauge, the denser the yarn. Stitches per inch indicate the number of horizontal rows of tufts per one inch of length. The more stitches per inch, the denser the carpet.

Density Factors (Woven Carpet):   This aspect is defined in terms of pitch and rows per inch. Pitch is the same as gauge for tufted carpet except that it is traditionally measured in a 27-inch width of finished goods. For example, if pitch is specified as 216, there are 216 rows of yarn in every 27 inches of width or the tufts are 1/8 inch apart. Rows per inch for woven carpet is the same as stitches per inch in tufted carpet.

CARPET CUSHION

Carpet cushion is defined as any material placed under carpet to provide softness when it is walked on. Other terms used in lieu of cushion include padding, underlayment, and lining. Carpet cushion under carpet offers several advantages as follows:

1. Adds additional acoustical value.
2. Adds additional thermal insulation.
3. Provides additional comfort.
4. Adds to the life of the carpet.

Carpets are available both with and without attached cushions. Cushion-backed carpets are much more difficult to seam, since they cannot be stretched to meet installation conditions.

Types of Cushions

Several types of carpet cushion are available, including:

Felt: Composed of natural hair, fiber, or a combination of both.

Foam and Sponge Rubber: Composed of synthetic rubber, latex, or urethane foams.

Proprietary Types: Including polyester pneumatic cellular fiber and resinated synthetic fiber.

Sizes and Weights

Cushion material comes in a variety of sizes and weights. Felt cushion is available in widths up to 12 feet and in weights from 32 to 86 ounces per square yard. Sponge rubber is available in widths up to 12 feet and in weights from 41 to 120 ounces per square yard.

PERFORMANCE CHARACTERISTICS

Flammability

Architects are cautioned to follow building codes in ascertaining which flame spread test method will be accepted for carpet. ASTM E84 is still required by many code authorities. A radient panel flooring flammability test, ASTM E648, is finding some favor as a measure of flame spread. Another test often cited by carpet producers is the Methanamine Tablet Test DOC FF1-70. Howver, since there is very little correlation between these test methods, users are advised to check with the local building code and the authorities.

Static Control

Static electricity is generated by friction when one walks across a carpet. The variables include humidity, shoe soles, and generic fiber types. Static can be controlled by raising the humidity above 40% or by the use of built-in static inhibitors such as special fibers to reduce the amount of static accumulation.

Test Methods

Test methods to ascertain specific characteristics with respect to colorfastness, soiling and appearance are as follows:

TESTS FOR CARPET CHARACTERISTICS

Colorfastness

Light	AATCC[a] 16-E
Crocking[b]	AATCC 8
Water	AATCC 107
Ozone	AATCC 129

Soiling	
Accelerated Soiling	AATCC 123
Service Soiling	AATCC 122
Visual Rating	AATCC 121

Appearance	
Shrinkage	ASTM D138
Abrasion	ASTM D1175
Moth & Larvae Resistance	ASTM D116
Mildew Resistance	Fed. Spec. CCC-T-191B

[a]AATCC American Association of Textile Chemists and Colorists.
[b]Crocking Loss of color due to rubbing off as a result of improper dye penetration.

CARPET INSTALLATION

Two major forms of installation are available: glue-down and tackless.

Glue-Down: Using no cushion or an attached cushion, carpet is glued down in place directly over a suitable substrate.

Tackless: Carpet stripping consisting of water-resistant plywood with angular pins protruding from the top is nailed or glued around the room perimeter. Carpet cushion is then placed within the confines of the carpet stripping. The carpet is then placed over the cushion and stripping by kicking or power stretching the carpet over the stripping.

Shop drawings should be required to show seam locations and pattern direction.

PAINTS AND COATINGS

The chemistry of paints and coatings is rather complex and is outside the needs of the individual responsible for selecting a paint or coating system for covering a specific surface in an architectural application. The architect's interests would be better served by acquiring a knowledge of the properties of paint products and their intended use in a particular environment on a specific substrate. This information is far more valuable to the architect or specifier than a knowledge of the chemistry of paints and coatings. These products are already chemically formulated, and the specifier need only know their properties and the needs of the project to make the proper selections.

NEED FOR PAINTING

Paints and coating systems are generally selected to provide one or more of the following requirements:

1. Protection, especially corrosion resistance
2. Appearance
3. Sanitation and cleanliness
4. Illumination and visibility
5. Safety and efficiency
6. Fire protection

Protection

Protection of the surface is afforded by a paint or coating system against moisture, chemical and in-dustrial fumes, sunlight, abrasion, dust and dirt, and temperature variations.

Appearance

Color, texture, and luster are properties that an architect can use in selecting a paint or coating system to enhance the appearance of specific surfaces and to create comfortable living and working areas.

Sanitation and Cleanliness

Specific paint and coating formulations may be selected to provide tile-like surfaces in areas of food preparation, food processing, shower and hydrotherapy areas, and similar spaces.

Illumination and Visibility

To brighten rooms and increase visibility, white and light-tinted paints are used to enhance these characteristics. In addition the degree of gloss used has an effect on the amount of reflected light and on the diffusion and distribution of light.

Safety and Psychological Impact

Colors are useful in (1) the degree of safety that can be improved around operating machinery; (2) identifying piping and ductwork; (3) identifying hazardous areas; (4) traffic control; (5) providing for the well-being of patients in hospitals by careful choice of colors for specific responses; (6) providing eye relief in areas where severe visual tasks are undertaken.

Fire Resistance

Selection of specific formulations with fire-resistive and fire-retardant properties is beneficial for application on specific substrates in certain areas requiring these properties.

PAINT MATERIALS AND THEIR PROPERTIES

There is no one universal paint material that can be applied to cover all substrates under all conditions of use. Paint formulations therefore are compromises by which manufacturers incorporate certain ingredients having specific properties to serve certain intended purposes over a period of time.

What Is Paint?

A paint or coating in a general sense is any liquid material, which when spread in a thin layer, solidifies into a film that obscures the surfaces on which it is applied, and provides a protective and decorative coating.

Paint Composition

Pigmented paints or coatings contain two basic materials—a pigment and a vehicle. The vehicle, which is the liquid portion, generally consists of two parts—one nonvolatile (the binder), and the other volatile (the solvent). The binder or nonvolatile material forms the film and derives its name from the fact that it binds or holds the pigment to the surface. The volatile part, the solvent, dissolves the film-forming material and is used to adjust the viscosity of the mixture to facilitate application. Each of these major components of paint may be composed of countless ingredients, each imparting different properties.

BINDERS

The different properties of paint and coatings are primarily the result of the different characteristics of the binders. The principal binders used both for architectural applications and where special problem application areas are involved are as follows:

Alkyd

Alkyd binders are oil-modified phthalate resins which dry by exposure to oxygen in the air. They are the most widely used because of their versatility and cost. They can be used both indoors and outdoors in various degrees of luster including flat, semigloss, and gloss. They should not be applied directly over fresh concrete, masonry, or plaster. Alkyds are a good decorative architectural coating for nonproblem areas.

Epoxy

Two types of epoxy are available—an epoxy ester and a two-component epoxy resin utilizing a polyamide or a polyamine hardener. The epoxy ester is similar to an alkyd and can be used indoors and outdoors. The epoxy ester is somewhat more resistant to chemical attack and fumes than alkyds. The two component epoxy is an expensive paint formulation intended for heavy duty service in chemical and industrial environments. The cured film has outstanding hardness, adhesion, and abrasion resistance and is also used as a tile-like glaze coating over concrete and masonry.

Epoxy Coal-Tar

Epoxy coal-tar is a two-component epoxy modified by the introduction of coal tar, intended for wet or submerged areas and for protection against splash and spillage of a wide variety of chemicals. It is limited in colors because of the coal-tar.

Inorganic

Inorganic binders consist of silicates of sodium, potassium, lithium and ethyl, and are used with "zinc-rich" anticorrosive paints for use on metal surfaces. They are extremely resistant to wet, humid and marine environments.

Latex

Latex binders constitute the most important and the most used water-based emulsion paints. Three latex polymers are used: polyvinyl acetate, acrylic, and butadiene-styrene. They have little odor and are very fast drying. Latex paints are used extensively for interior wallboard and plaster, and on exterior masonry and primed wood.

Oil

Linseed oil is one of the chief binders in exterior house paints and for primers for exterior structural steel. The oil house paints have the longest history of performance of modern-day paints. Since they have good wetting properties, they are used as primers for structural steel since surface preparation is less demanding. Although they are not particularly hard or resistant to abrasion, chemicals, or industrial fumes, they are durable in normal environments.

Oil Alkyd

Oil alkyd binders are a combination of linseed oil and alkyd resin binders, blended to provide the inherent good qualities of both. Oil alkyd has improved hardness, is faster drying, wets well, and has good gloss retention.

Oleoresinous

Oleoresinous binders have been processed by combining drying oils and hard resins by a cooking process. The resin increases the hardness and chemical resistance as compared to a straight oil paint. Depending on the resin used, there may be increased gloss, faster drying, improved adhesion, and greater durability.

Phenolic

Phenolic binders are among the first truly synthetic resins. They may be used as clear finishes or pigmented in a range of colors. Phenolic paints are used as topcoats on metal for humid environments and as primers for fresh water immersion.

Rubber-Base

Cholorinated-rubber-base binders are solvent thinned and are not to be confused with the latex types, which are water thinned. Chlorinated rubber is a product of chlorine and polyisoprene. It has outstanding resistance to water and common corrosive chemicals, and possesses a high degree of impermeability to water vapor. It is highly resistant to strong acids and alkalis. It is used for areas where excessive moisture exists, such as swimming pool areas, hydrotherapy areas, wash and shower rooms, commercial kitchens, and laundry rooms.

Silicone Alkyds

Silicone alkyd binders are a combination of silicone and alkyd resins. It has excellent color and gloss retention and is very suitable for exterior architectural steel surfaces. It is also resistant to high temperatures and is used on metal chimneys.

Urethane

Three types of urethane finishes are available. *Single-component types* are available in two formulations—one an *oil-modified urethane* which has been modified with drying oils and alkyds, and the other a *moisture-curing urethane* which cures by solvent evaporation and reaction with moisture in the air. The *two-component system* has outstanding abrasion and chemical resistance, hardness, flexibility, and good exterior gloss and color retention. The oil-modified urethane is similar to the phenolic varnishes, has better initial color and color retention and can be used as an exterior spar varnish or tough interior floor finish. The moisture-curing urethane has excellent flexibility and chemical and water resistance.

Vinyls

Vinyls are copolymers of vinyl chloride and vinyl acetate. They are low in solids and multiple coats are required. They have good weathering qualities and are used on exterior metal.

TABLE 9-2
COMPARISON OF BINDERS ON SUBSTRATE AND ENVIRONMENT

	Alkyd	Epoxy	Latex	Oil	Phenolic	Chlorinated Rubber	Urethane	Vinyl
Substrate								
Concrete	NR	VG	VG	NR	NR	VG	G	VG
Metal	VG	VG	F	VG	VG	G	G	EX
Wood	G	G	G	G	G	NR	G	NR
Environment								
Industrial	F	EX	NR	NR	G	G	VG	EX
Marine	F	EX	F	NR	G	EX	EX	EX
Rural	VG	EX	VG	G	G	G	VG	EX

EX—excellent. VG—very good. G—good. F—fair. NR—not recommended.

Binder Comparisons

Paint formulations based on current chemical technology are so ever changing that the specifier and user are urged to compare products of competing manufacturers carefully and to stay abreast of these developments since there is a constant upgrading of products and development of new binders and resins.

For a comparison of binders with reference to specific substrates and environment see Table 9-2. Also see Table 9-3 for performance characteristics.

SPECIAL PURPOSE PAINTS AND COATINGS

For certain specific requirements, there are special paints and coatings that have been formulated to cope with these problems as follows:

TABLE 9-3
PRINCIPAL PROPERTIES OF PAINT BINDERS

Property	Alkyd	Epoxy	Latex	Oil	Phenolic	Chlorinated Rubber	Urethane	Vinyl
Adhesion	VG	EX	G	VG	G	G	VG	F
Flexibility	G	EX	G	EX	G	G	EX	EX
Hardness	G	EX	G	F	VG	VG	EX	G
Resistance to								
Abrasion	G	EX	G	P	EX	G	EX	VG
Acid	F	G	G	P	VG	EX	EX	EX
Alkali	F	EX	G	P	G	EX	EX	EX
Detergent	F	EX	G	F	VG	EX	EX	EX
Heat	G	G	G	F	G	VG	G	P
Strong Solvents	P	EX	G	P	G	P	EX	F
Water	F	G	F	F	EX	EX	EX	EX

EX—excellent. VG—very good. G—good. F—fair. NR—not recommended.

Abrasion-Resistance

To resist abrasion for traffic areas and to withstand repeated washings and scrubbings as in institutional usage (e.g., hospitals, schools, food-processing), there are several paints that are especially designed and formulated for this purpose. Those include epoxy coatings, polyurethane coatings, and polyester-epoxy coatings. Tilelike coatings for walls are produced to meet Fed. Spec. TT-C-550.

Fire Protective

Two aspects of fire protection should be understood— one dealing with fire retardancy, the other with fire resistance.

Fire-Retardant Paints

Fire-retardant paints are used to inhibit the spread of fire on combustible surfaces such as wood. Typically they are used to obtain a Class A flamespread of between 0-25 as per ASTM E84. Interior type paints having this capability should be specified to meet Fed. Spec. TT-P-26. Exterior paints should be specified to meet Military Specification MIL-C-46081.

Fire-Resistant Paints

Fire-resistant paints are those which are used to obtain UL rated protection of steel framing of 1 or 2 hours based on ASTM E119 fire test. The coatings which provide this protection are either intumescent mastics or subliming mastics.

Heat Resistance

Where heated surfaces such as high-pressure steam, metal chimneys, or equipment are to be painted, heat-resistive paints meeting Fed. Spec. TT-P-28 or TT-E-496 may be utilized.

Mildew Resistance

Where mildew may be a problem, paints containing zinc oxide such as Fed. Spec. TT-P-102 and TT-P-105 may be used, or a mildewcide may be added to the paint. Mildew resistance may be tested in accordance with Federal Test Standard No. 141, Method 6271.1.

Slippage Resistance

Nonslip coatings are made by incorporating an abrasive grit into the paint formulation, or, it may be added to the paint or broadcast onto the freshly painted surface.

PAINTING

SURFACE PREPARATION

The importance of proper surface preparation to the durability and longevity of a coating system cannot be overemphasized. Without this preparation the most costly paint material applied in the most professional manner will not live up to its expectations. Thorough cleaning of the surface, and pretreatment where needed, will enhance adhesion of the coating and in the case of metals provide a barrier against corrosion.

Methods of surface preparation may vary from a light cleaning or brushing to a heavy sandblasting for removal of dirt, grit, or scale, and the use of solvent washes to remove oils or grease.

Steel Surfaces

Solvent Cleaning

A very effective method to remove oil, grease, waxes, and other solvent-soluble materials from steel is to employ a system described in Steel Structures Painting Council, SSPC SP1. Solvent cleaning precedes mechanical treatment.

Mechanical Treatment

Three methods may be employed to remove rust, mill scale, dirt and dirt incrustations. The degree to which these will be removed depends upon the specific treatment selected as follows:

1. *Hand Cleaning.* Wire brushes and scrapers are used to remove loose rust and loose scale. This process will not remove heavy or tightly bound rust and mill scale. It is used where mild, noncorrosive atmospheres will be encountered and a linseed oil wetting type of primer will be used. The process can be specified to meet SSPC-SP2.

2. *Power Tool Cleaning.* This method is obviously much faster than hand tool cleaning and is accomplished with power wire brushes, power sanders, power grinders, or by a combination of these tools. It may remove small amounts of tightly adhering contaminants which hand tools may not remove. This process can be specified to meet SSPC-SP3.

3. *Blast Cleaning.* Several methods utilizing blast cleaning are available in which sand, synthetic grit, or other abrasive materials are utilized to remove rust, scale, and other contaminants. The degree to which these materials are removed are contained in Steel Structures Painting Council specifications identified with the method.

　　a. *White Metal Blast—SSPC-SP5.* This is the ulitmate, most expensive and effective method. It removes all rust, mill scale and all other contaminants, leaving a completely clean uniform surface with an ideally roughened textured surface for maximum paint adhesion and durability under the most severe conditions. This method is warranted only for the most demanding service in corrosive environments. The paint coating system should be a "zinc-rich" system, an epoxy system or a high build vinyl system, depending on the exposure.

　　b. *Near-White Metal Blast—SSPC-SP10.* In this method a small specified amount of streaking and shadowing will appear across the general surface area. It is less expensive than a white blast and is generally used for exposed architectural steel utilizing coating systems of "zinc-rich" primers, and either silicone alkyd or high-build vinyl finishes, depending on the environment.

　　c. *Commercial Blast—SSPC-SP6.* This method will remove loose rust and scale resulting in a satisfactory surface that is generally adquate for all but the most vigorous types of service exposure. The coating system to be selected should suit the environment except the most rigorous and corrosive.

　　d. *Brush-Off Blast—SSPC-SP7.* This method of blast cleaning is intended to remove only loose rust scale leaving only tightly adhering, intact mill scale and rust. This degree of surface preparation is comparable or superior to hand or power tool cleaning and should be used only for mild exposures with linseed oil primers having good wetting ability.

Galvanized Steel

Adhesion of paint systems to new galvanized steel is especially difficult. The paint peels or flakes off often after a short period of exposure. Oil, grease and protective temporary oils should be removed by solvent cleaning. A pretreatment wash primar (MIL-C-15328) should then be applied to develop good adhesion for a zinc-dust primer, Fed. Spec. TT-P-641. Wash primers are thin coats containing polyvinyl butyral resin, phosphoric acid, and a rust-inhibitive pigment.

Aluminum

When aluminum is to be painted, it is best prepared by solvent cleaning and the application of a wash primer as described above for galvanized steel surfaces.

Wood

Wood surfaces should be clean, free of cracks and splinters, and have a moisture content below 15% for exterior wood and 10% for interior wood. Cracks and nail holes should be filled with putty or plastic wood. Puttied areas should then be sanded smooth. Knots, pitch streaks or visible sap spots should be treated with Formula WP-578 Knot Sealer developed by the Western Pine Association.

Concrete and Masonry

Surfaces should be clean, free of dust, dirt, oil, grease, efflorescence, chalk, and loose material. Aging of the surfaces permits them to dry out and neutralizes the alkalinity. Surfaces are cleaned by bristle brushing or hosing down with water. Scraping, wire brushing or sandblasting may be employed on concrete where necessary. Efflorescence may be removed from concrete by wire brushing or sand blasting. Large cracks, holes and other blemishes should be repaired by patching with cementitious mixtures. Open textured masonry may be filled with a sand cement grout coat if a smooth surface is desired. Concrete surfaces that have a glazed finish resulting from smooth nonabsorbent forms should be etched with muriatic acid (5 to 10% solution hydrochloric acid.)

Plaster

Cracks, holes, indentations and similar defects should be repaired with a spackling compound, or patching plaster, and then sanded smooth. New plaster should be aged a minimum of 2 weeks before application of latex paints. If oil or oleoresinous paints are used, new plaster should age at least 2 to 3 months before paint application.

Gypsum Drywall

Minor cracks and holes should be repaired with finishing compound and then sanded when dry. Surfaces should then be wiped free of dust, preferably with a damp sponge.

PAINT SELECTION

How does the user or specifier go about making decisions concerning the selection of a paint or coating system? By establishing a rationale for the selection, the user can zero in on the paint or coating system that will best suit the intended purpose.

Type of Exposure

Questions concerning exposure can be related to whether it is an interior or exterior environment and the conditions of that environment as follows:

Interior	Exterior
Chemical	Dry
Dry	Industrial
Industrial	Marine
Wet	Rural
	Temperature extremes
	Urban
	Wet

Surfaces and Substrates

Surfaces to be painted in the same space may or may not require the same type of paint based on a number of factors. Ceilings require paints with good reflectivity for optimum illumination. Walls require a paint that will not run or sag on a vertical surface when applied, and one which will permit repeated washings for maintenance purposes. Floors require a paint that will withstand abrasion and will dry quickly so that the space may be utilized without much downtime.

Substrates such as drywall, plaster, concrete, concrete block, wood and metal must be reviewed for their special needs and their special environments. Some woods require special primers or sealers to prevent resins in knots from bleeding through the paint. Plaster and concrete may be excessively alkaline when new and may require pretreatment before application of oil paints. Concrete floors may have a glaze which requires etching to insure paint adhesion. Some metals, such as aluminum and galvanized steel, may require wash primers to etch the surface to insure adhesion of coating systems.

By reviewing Tables 9-2 and 9-3 together with manufacturers' literature, the user and specifier will be in a better position to make evaluations and selections.

BIBLIOGRAPHY

CHAPTER 1—PERFORMANCE CONSIDERATIONS

Canadian Building Digests, National Research Council of Canada, Ottawa, Canada.

Construction Materials Evaluation & Selection. Harold J. Rosen and Philip M. Bennett, eds. John Wiley & Sons, New York, 1979.

Testing Building Constructions and the Performance Concept. Robert F. Legget, Division of Building Research Paper No. 701, Oct. 1976 National Research Council of Canada, Ottawa, Canada.

CHAPTER 2—SITEWORK

Finishing Concrete Slabs, Exposed Aggregate, Patterns & Colors. Portland Cement Association, 1979.

CHAPTER 3— ARCHITECTURAL CONCRETE

Architectural Precast Concrete. Precast Concrete Institute, 1973.

Bushhammering of Concrete Surfaces. Portland Cement Association, 1972.

Canadian Building Digests Nos. 15, 56, 93, 103, 116, 136 and 203, National Research Council, Ottawa, Canada.

Color, Form & Texture in Architectural Precast Concrete. Precast Concrete Institute, Chicago.

Color & Texture in Architectural Concrete. Portland Cement Association 1980.

Fabrication, Handling and Erection of Precast Concrete Wall Panels. ACI Journal, April 1970.

Finishes to In-Site Concrete. J. Gilchrist Wilson, Exposed Concrete Finishes, Vol. 1, John Wiley & Sons, New York, 1979.

Finishes to Precast Concrete. J. Gilchrist Wilson, Exposed Concrete Finishes, Vol. 2. John Wiley & Sons, New York, 1964.

Finishing Concrete Slabs, Exposed Aggregate, Patterns & Colors. Portland Cement Association, 1979.

Guide to Cast-in-Place Architectural Concrete Practice ACI Journal, July 1974.

Jobsite Precast Concrete Panels—Textures, Patterns & Designs. Portland Cement Association, 1975.

Plywood for Concrete Forming. American Plywood Association, 1971.

Site Cast Architectural Concrete. Concrete Construction, November 1972, Construction Publications Inc., Elmhurst, IL.

Tilt-Up Concrete Walls. Portland Cement Association, 1970.

CHAPTER 4—MASONRY

Brick

Brick & Tile Engineering. Harry Plummer, Brick Institute of America, 1962.

Canadian Building Digests Nos. 2, 6, 21, 169 and 185. National Research Council, Ottawa, Canada

Details from Brick in Architecture. Brick Institute of America, 1978.

Technical Notes on Brick Construction, Brick Institute of America

Concrete Block

Concrete Masonry Handbook. Portland Cement Association, 1976.

TEK Bulletins. National Concrete Masonry Association, Herndon, VA.

Stone

Indiana Limestone Handbook. Indiana Limestone Institute of America, Inc.

Marble Design Manual II. Marble Institute of America. Farmington, MI.

Stone Catalog. Building Stone Institute, New York.

CHAPTER 5—METALS

Aluminum

A Guide to Aluminum Extrusions. The Aluminum Association, 1979.

Aluminum Standards and Data. The Aluminum Association, 1982.

Designation System for Aluminum Finishes. The Aluminum Association, 1976.

Copper, Brass, and Bronze

Copper Brass Bronze Design Handbook. Copper Development Association, Greenwich, CT.

Stainless Steel

Stainless Steel: Concepts in Design and Fabrication. American Iron & Steel Institute.

Steel

Fire Safe Structural Steel: A Design Guide. American Iron & Steel Institute.

General

Metal Finishes Manual. National Association of Architectural Metal Manufacturers.

CHAPTER 6— ARCHITECTURAL WOODWORK

Architectural Woodwork Quality Standards. Architectural Woodwork Institute

Architectural Woodwork Interiors—Wall & Ceiling Treatment. Architectural Woodwork Institute

Building Code Flame Spread Classifications. Architectural Woodwork Institute.

Factory Finishing of Architectural Woodwork. Architectural Woodwork Institute.

Fine Hardwoods. American Walnut Association.

Fine Hardwoods Selectorama. American Walnut Association.

Fine Hardwood Veneers for Architectural Interiors. American Walnut Association.

Guide to Wood Species. Architectural Woodwork Institute.

High Pressure Laminates as an Architectural Woodwork Material. Architectural Woodwork Institute.

Small Homes Council. University of Illinois D7. 2 Plywood.

Structural Glued Laminated Timber, American Institute of Timber Construction, Spec-Data sheet, 1980.

Wood, Colors & Kinds. U. S. Forest Products Laboratory, Argricultural Handbook No. 101, 1956.

Wood Handbook, U.S. Forest Products Laboratory.

CHAPTER 7—THERMAL AND MOISTURE PROTECTION

Built-Up Roofing

Canadian Building Digests Nos. 24, 67, 68, 73, 74, 95, 99, 150, 176, 179, 181 and 211. National Research Council, Ottawa, Canada.

Manual of Built-Up Roof Systems. C. W. Griffin, McGraw-Hill, New York.

NRCA Roofing & Waterproofing Manual. National Roofing Contractors Association.

Single-Ply Roofing

1984 Handbook of Single Ply Roofing Systems, Harcourt Brace Jovanovich, Cleveland.

Single Play Roofing Membrane, National Roofing Contractors Association.

Single Ply Roofing Technology, ASTM STP 790, June 1981.

Sheet Metal

Architectural Sheet Metal Manual. Sheet Metal & Air Conditioning Contractors National Association (SMACNA), 1979.

Architectural Sheet Metal. SMACNA, April 1976.

Aluminum Sheet Metal Work in Building Construction. Aluminum Association, Sept 1971.

Contemporary Copper. Copper Development Association.

Stainless Steel: Suggested Practices for Roofing, Flashing. American Iron & Steel Institute, Nov. 1972.

Thermal Insulation

An Assessment of Thermal Insulation Materials & Systems for Building Applications. U.S. Dept of Energy. BNL-50862, June 1978.

Canadian Building Digests Nos. 16, 36, 52, 70, 102, 167 and 178. National Research Council of Canada, Ottawa, Canada.

Commerical & Industrial Insulation Standards. Midwest Insulation Contractors Association.

Joint Sealing Materials

Canadian Building Digests Nos. 19, 96, 97, 155, and 158. National Research Council of Canada, Ottawa, Canada.

Chemical Materials for Construction. Philip Maslow, Structures Publishing Co., Farmington, MI.

Construction Sealants & Adhesives. Julian R. Panek and John R. Cook, John Wiley & Sons Inc., New York, 1984.

Guide to Joint Sealants for Concrete Structures. ACI Committee 504R, 1977.

CHAPTER 8—GLASS AND CURTAIN WALLS

Glass

Canadian Building Digests Nos. 55, 60, 101 and 132. National Research Council, Ottawa, Canada.

Glazing Manual 1980. Flat Glass Marketing Association, Topeka, KA.

Installation Recommendation, Tinted & Reflective Glass. TSR 130, PPG Industries, Pittsburgh, PA.

Installation Recommendations for Twindow (Insulating Glass). TSR 230, PPG Industries, Pittsburgh, PA.

Sealant Technology in Glazing Systems, STP 638, ASTM 1977.

Sloped Glazing Guidelines AAMA TIR-A7-83, Architectural Aluminum Manufacturing Association, Chicago, IL.

Curtain Walls

Aluminum Curtain Walls Series, 1-12 Architectural Aluminum Manufacturers Association (AAMA), Chicago, IL.

Canadian Building Digests Nos. 28, 34, 39, 40, and 48, National Research Council, Ottawa, Canada.

Design Wind Loads for Aluminum Curtain Walls. TIR-A2-1975, AAMA.

Metal Curtain Wall, Window, Store Front & Entrance Guide Specifications Manual, AAMA, 1976.

Methods of Tests for Curtain Walls. 501–83, AAMA.

Minimum Design Loads for Buildings and Other Structures, Section 6, Wind Loads. A 58.1-1982, ANSI.

Window & Wall Testing. STP 552, ASTM, 1972.

CHAPTER 9—FINISHES

Carpeting

Carpet Specifiers Handbook 1974, The Carpet & Rug Institute, Dalton, GA.

Contract Carpeting. Lila Shoshkes, Watson-Guptil Publications 1974, New York.

Ceramic Tile

Canadian Building Digest No. 206, National Research Council, Ottawa, Canada.

Ceramic Tile Manual, 1982, Ceramic Tile Institute, Los Angeles.

Handbook for Ceramic Tile Installation, Tile Council of America, Princeton, NJ.

Concrete Floors

Canadian Building Digest No. 22, National Research Council, Ottawa, Canada.

Finishing Concrete Slabs, Exposed Aggregate, Patterns & Colors, Portland Cement Association.

Guide for Concrete Floor & Slab Construction, ACI Committee 302, 1980, American Concrete Institute.

Terrazzo

Terrazzo Technical Data, The National Terazzo & Mosaic Association, Des Plaines, IL.

Paints and Coatings

Canadian Building Digests Nos. 76, 78, 79, 90, 91, 98 and 131. National Research Council, Ottawa, Canada.

Organic Coatings, 1968, A.G. Roberts NBS BSS 7, National Bureau of Standards.

Paints & Protective Coatings, Depts. of the Army, Navy and Air Force, 1969. U.S. Government Printing Office.

Steel Structures Painting Manual, 1982, Vol. 1 Good Painting Practices; Vol. 2, Systems & Specifications, Steel Structures Painting Council, Pittsburgh, PA.

INDEX